FORTSCHRITTE DER PHYSIK

Volume 33 · 1985
Number 1

PROGRESS OF PHYSICS

CONTENTS:

AKADEMIE-VERLAG · BERLIN

ISSN 0015-8208 Fortschr. Phys., Berlin 33 (1985) 1, 1–55 EVP 10,— M

Instructions to Authors

1. Only papers not published and not submitted for publication elsewhere will be accepted.
2. Manuscripts should be submitted in English, with an abstract in English. Two copies are desired.
3. Manuscripts should be no less than 30 and preferably no more than about 100 pages in length.
4. All manuscripts should be typewritten on one side only, double-spaced and with a margin 4 cm wide. Manuscript sheets should be numerated consecutively from "1" onwards. Footnotes should be avoided.
5. The titel of the paper should be followed by the author's name (with first name abbreviated), by the institution and its address from which the manuscript originates.
6. Figures and tables should be restricted to the minimum needed to clarify the text. They should be numbered consecutively and must be referred too in the text and on the margin.
   Figures and tables should be added to the manuscript on separate, consecutively numerated sheets. The tables should have a headline. Legends of figures should be submitted on a separate sheet. All figures should bear the author's name and number of figure overleaf.
   Photographs for half-tone reproduction should be in the form of highly glazed prints.
   Line drawings should be in a form suitable for reproduction. The lettering should be sufficiently large and bold to permit reduction. If requested, original drawings and photographs will be returned to the author upon publication of the paper.
7. Formulae should not be written to small and not with pencil. Separate lines for formulae are desirable. SI-units should be used.
   Letters in formulae are normally printed in italics, numbers in ordinary uprigth typeface.
   Underlining to denote special typefaces should be done in accordance with the following code:
   Italics: wavy underlined with pencil (only necessary for type written symbols in the text)
   Boldface italics (vectors): straightforward and wavy underlined
   Upright letters (all abbreviations like all units (cm, g, ...), all elements and particles (H, He, ..., n, p, ...), elementary mathematical functions like Re, Im, sin, cos, exp, ...): black underlined
   Greek letters: red underlined
   Boldface Greek letters: red underlined twice
   Uprigth Greek letters (symbols of elementary particles): red and black underlined
   Large letters: underlined with pencil twice
   Small letters: overlined with pencil twice
   (This will be necessary for handwritten letters that do not differ in shape, as *c C*, *k K*, *o O*, *p P*, *s S*, *u U*, *v V*, *w W*, *x X*, *y Y*, *z Z*).
   It will help the printer if position of subscripts and superscripts is marked with pencil in the following way: $a_{\uparrow}$, $b^{i}_{\downarrow}$, $M_{\uparrow}{}^{j}_{\downarrow}$, $M_{\hat{i}j}$, $W_{n\hat{1}}$
   Please differentiate between following symbols: a, α; α, $\propto$, ∞; α, d; c, C, $\subset$; e, l, $\mathcal{E}$, ε, $\in$; k, K, $\varkappa$; $\varkappa$, x, X, χ, $\times$; l, 1; o, O, σ, 0; p, ϱ; u, U; $\cup$, v, ν, $\vee$; φ, ϕ, Φ, $\emptyset$; ϑ, θ Θ.
8. Each paper should be followed by a list of references with consecutive numeration. The numbers should be placed on the line between squared brackets or typewritten inclined lines. Articles in journals should be cited with author's name and abbreviated first name, title of journal, volume number (underlined), year of publication (in brackets) and page number. In case of books authors' name and abbreviated first name, title, place, and year of publication should be given.
9. Manuscripts should be submitted fit for printing; badly arranged manuscripts are returned.
10. Proof reading should be limited to the correction of typographical errors. Deletions and insertions are not permitted in proof reading.
11. Of every paper published the author(s) will receive 30 reprints free of charge.
12. Within the framework of legal protection the publishers have the right of publication, distribution and translation reserved. Without express permission it is not allowed to produce photocopies, microfilms etc. of this journal or parts of it.

FORTSCHRITTE DER PHYSIK

PROGRESS OF PHYSICS

Volume 33 · 1985

AKADEMIE-VERLAG · BERLIN

ISSN 0015-8208 Fortschr. Phys., Berlin 33 (1985)

The journal "Fortschritte der Physik/Progress of Physics" is devoted to the theoretical and experimental study of the fundamental constituents of matter and their interactions, so to elementary particle physics, classical and quantum field theory, theory of gravitation, thermodynamics and statistics, nuclear physics, laser physics, and plasma physics.
The articles should have in general review character.

Manuscripts should be addressed to the Editor-in-Chief of the journal

Prof. Dr. F. Kaschluhn
Sektion Physik der Humboldt-Universität zu Berlin
DDR-1086 Berlin, PSF 1297

or to editors or to members of the Advisory Board.

Terms of subscription for the journal "Fortschritte der Physik/Progress of Physics"

Orders can be sent

- *in the GDR:* to Deutsche Post, Zentralvertrieb des PZV (B), 7930 Herzberg/Elster or to the Akademie-Verlag Berlin, DDR-1086 Berlin, Leipziger Str. 3–4, PF-Nr. 1233;
- *in the other socialist countries:* to a book-shop for foreign language literature or to the competent news-distributing agency;
- *in the FRG and Berlin (West):* to a book-shop or to the wholesale distributing agency Kunst und Wissen, Erich Bieber OHG, Wilhelmstr. 4–6, D-7000 Stuttgart 1;
- *in the other Western European Countries:* to Kunst und Wissen, Erich Bieber GmbH, Dufourstr. 51, CH-8008 Zürich;
- *in other countries:* to the international book- and journal-selling trade, to Buchexport, Volkseigener Außenhandelsbetrieb der DDR, Postfach 160, DDR-7010 Leipzig, or to the Akademie-Verlag Berlin, DDR-1086 Berlin, Leipziger Str. 3–4, PF-Nr. 1233.

Zeitschrift „Fortschritte der Physik/Progress of Physics"

Herausgeber: Prof. Dr. F. Kaschluhn, Prof. A. Lösche, Prof. Dr. R. Rompe.

Verlag: Akademie-Verlag Berlin, DDR-1086 Berlin, Leipziger Str. 3–4;
Fernruf: 2236201 und 2236229; Telex-Nr.: 114420; Bank: Staatsbank der DDR, Berlin, Konto-Nr.: 6836-26-20712.

Chefredakteur: Dr. Lutz Rothkirch.
Anschrift der Redaktion: Sektion Physik der Humboldt-Universität zu Berlin, DDR-1040 Berlin, Invalidenstraße 110.

Veröffentlicht unter der Lizenznummer 1324 des Presseamtes beim Vorsitzenden des Ministerrates der Deutschen Demokratischen Republik.

Gesamtherstellung: VEB Druckhaus „Maxim Gorki",
DDR-7400 Altenburg, Carl-von-Ossietzky-Straße 30/31.

Erscheinungsweise: Die Zeitschrift „Fortschritte der Physik" erscheint monatlich. Die 12 Hefte eines Jahres bilden einen Band. Bezugspreis je Band 216,– DM zuzüglich Versandspesen. Preis je Heft 18,– DM. Der gültige Jahresbezugspreis für die DDR ist der Postzeitungsliste zu entnehmen.

Bestellnummer dieses Heftes: 1027/33.

AN (EDV) 57618

Contents of Volume 33

Number 1

Number 2

Number 3

Number 4

Number 5

Number 6

Number 7

Number 8

Number 9

Number 10

Number 11/12

VOLUME 33 1985 NUMBER 1

FORTSCHRITTE DER PHYSIK

Fortschr. Phys. 33 (1985) 1, 1–55

Beta Decay Far from Stability and Its Role in Nuclear Physics and Astrophysics[1])

HANS VOLKER KLAPDOR

Max-Planck-Institut für Kernphysik, Heidelberg, Federal Republic of Germany

Contents

1. Introduction

The importance of the shape of the beta strength function $S_\beta(E)$ in nuclear physics and astrophysics (Fig. 1) has made its investigation a challenging task, in particular, since the only description of the beta decay properties of nuclei far from stability existing for the last decade was based on very schematical assumptions of the shape of $S_\beta(E)$, and consequently the previous and many current extensive efforts to understand the properties of nuclides far from stability as well as the synthesis of heavy elements and other problems in astrophysics and nuclear physics were performed by means of an inappropriate description of the beta decay properties.

[1]) Invited review talk presented at the "Internat. School-Seminar on Heavy Ion Physics", Alushta, Crimea, USSR, 14.–21. April 1983 (updated version).

This paper gives a microscopic description of the beta decay of nuclei far from stability. By a TDA treatment and making use of the most recent predictions of fission barrier heights and of optical potentials from infinite nuclear matter calculations, a set of beta decay half-lives, β-delayed neutron and fission rates has been obtained for ~6000 neutron-rich nuclei between beta stability line and and neutron drip line (KLAPDOR et al. (1981, 1982, 1984); THIELEMANN et al. (1983)).

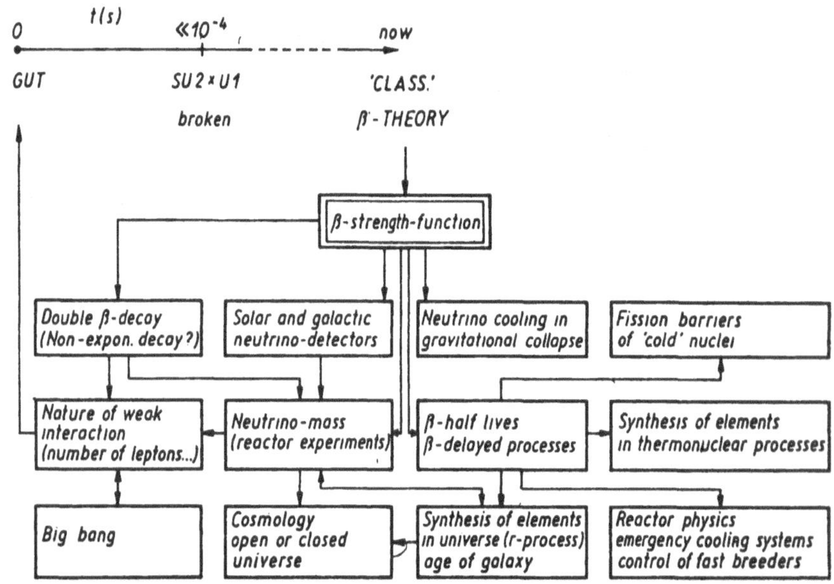

Fig. 1. Beta decay (the shape of the beta strength function) enters in a decisive way in various processes and problems in nuclear, particle and astrophysics and nuclear techniques

The large impact these results could have on various problems in nuclear physics and astrophysics we had considered qualitatively in earlier papers. In this report, the partly dramatic results of quantitative investigations can be presented (see also the reviews by KLAPDOR (1983a) and (1984a)).

As a consequence of the systematically shorter beta decay half-lives of neutron-rich nuclei compared to earlier estimates, a solution has been obtained (KLAPDOR et al. (1981)) for the problem of the astrophysical site of the *r*-process, which has been an open question in astrophysics for more than two decades.

It is found that the *r*-process is terminated by β-delayed fission before earlier discussed mechanisms come into play, so that as a consequence of the island of beta-delayed fission no superheavy nuclei will be produced in nature.

Cosmochronology by the heavy chronometers (^{129}I, ^{232}Th, ^{235,238}U, ^{244}Pu) produced exclusively in the *r*-process is put for the first time on a reliable basis by taking into account β-delayed fission in the *r*-process path and during the decay to the stability line. The age of the universe we find (THIELEMANN et al. (1983), KLAPDOR (1984), (1984a)) from these chronometers to be $(21.8^{+2}_{-4}) \cdot 10^9$a, which is supported by recent observations of ages of globular clusters.

A precise knowledge of the beta strength function is shown to be essential for determining the efficiency of detectors for solar and galactic neutrinos, in particular for any spectroscopic application.

The only experimental method allowing the investigation of fission barriers of "cold" nuclei far from the beta stability line — which is a challenge for probing the "contours" of the superheavy island without extending to the region of unknown elements (LAZAREW et al. (1980)) — requires a detailed knowledge of the beta strength function.

The long known "odd-even" reversal in thermonuclear explosions (DORN and HOFF (1965)) is understood as natural consequence of β-delayed fission in the actinide region. The island of β-delayed fission is further responsible for the fact that no nuclides with $A > 257$ have been observed in thermonuclear processes and its effect will have to be considered in proposals to produce heavy elements by several successive neutron bursts.

We find (GROTZ et al. (1983)) that the Gamow-Teller strength function can be sensitive to nucleon density distributions in nuclei, which opens a new field of application of (p, n) charge exchange reactions.

A calculation (GROTZ, KLAPDOR, METZINGER (1983a), KLAPDOR and GROTZ (1984)) of the 2ν double beta decay of $^{128,130}Te$ and ^{82}Se in the BCS pairing model including spin-isospin correlations and the quenching effect of ΔN^{-1} excitations shows that the often used Pontecorvo argument may lead to incorrect half-life ratios. This has severe impact on the discussion of neutrinoless ββ-decay and a possible neutrino mass.

Double beta decay has been discussed also (KLAPDOR and GROTZ (1983, 1984)) in connection with non-exponential decay, presently considered for the decay of the proton in connection with tests of grand unification theories (GUT's).

The accuracy with which the beta strength function of short-lived fission products can now be predicted, is of economic interest for optimizing the emergency cooling systems of light water reactors (KLAPDOR (1982, 1983)).

The antineutrino spectrum in the reactor core can, as the electron spectra, be calculated with an accuracy increased by almost an order of magnitude over calculations treating the beta decay in a more "schematical" way (KLAPDOR and METZINGER (1982)). These calculations are found to be decisive for the interpretation of reactor neutrino oscillation experiments presently under way at power reactors.

2. Calculation of Beta Decay Properties for Nuclei Far from Stability

Since nuclear spectroscopy of various types, the most important being isovector $M1$ γ-decay of isobaric analogous states (KLAPDOR (1971, 1976)) and (p, n) charge exchange reactions (DOERING et al. (1975), GOODMAN et al. (1980), GROTZ et al. (1983)) has led to a break-through in the understanding of beta decay in recent years (for a survey see KLAPDOR (1983a)) and since a better understanding has been reached of the quenching of $\beta^{\pm}$ strength by Δ excitations (GROTZ, KLAPDOR, METZINGER (1983b)), calculations of β decay have become possible with rather high accuracy. The necessary theoretical 'tools' exist partly already for some time. Following the work of FUJITA and IKEDA (1965, 1966) who extended the conventional pairing model by additional short or long range residual proton-neutron interactions specific to GT decay to reproduce the collective effects, HALBLEIB and SORENSEN (1967) developed an unlike-particle RPA for a model containing pairing and GT residual forces, thus adapting the RPA method for the study of isospin-flip processes. A careful investigation of the importance of various types of refinements (or neglections) in the calculation of GT β-strength functions in a shell model with pairing and inclusion of a residual, long range Gamow-Teller force has been made by RANDRUP (1973) and GROTZ and KLAPDOR (1983). A remarkable result is that the TDA solution (which results from the RPA by ignoring spin-isospin correlations) for the β^- strength function is found to be in general rather insensitive against various kinds of sophistication of the theory in contrast to the β^+ strength function which can be massively dependent on various types of g.s. correlations.

It is well-known that the RPA solutions can be interpreted as normal modes of the A-nucleon system around the Hartree-Fock ground state and that the common quasi-boson approximation is equivalent to the assumption of vibrations with small amplitudes. It is thus not surprising that the RPA is equivalent to a field approximation. The self-consistency condition for the oscillating average field proportional $\sigma\tau$ (which generates the coherent state) leads to a dispersion relation which is equivalent to the RPA equation for a separable force (BOHR and MOTTELSON (1975)).

2.1. The model for calculation of S_β

We have performed a calculation of the beta strength function $S_\beta(E)$ for neutron-rich nuclides up to the neutron drip line for $Z = 6-114$ using the Tamm-Dancoff approximation (KLAPDOR and ODA (1980), KLAPDOR et al. (1981, 1982)). The simplest type of GT force is used as residual interaction (as in RANDRUP (1973)), namely the (long range) spin-isospin flip part of the two-body interaction, $(\sigma \cdot \sigma)(\tau \cdot \tau)$, neglecting any radial or angular dependence. In some cases (sections 3.2.3. and 3.2.4.) refinements have been made, by including pairing and spin-isospin correlations and the excitation of ΔN^{-1} states (GROTZ, KLAPDOR, METZINGER (1983), KLAPDOR (1983b, 1984a)).

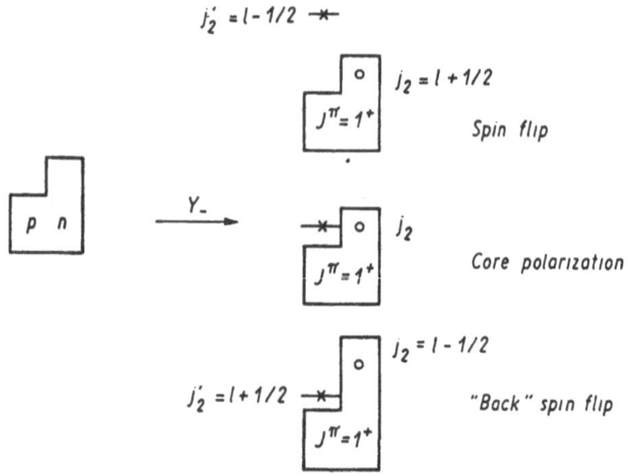

Fig. 2. The NN^{-1} particle-hole states excited by the Gamow-Teller operator (schematic)

As basis states in the daughter nucleus we take those configurations generated by GT single particle transitions, i.e. core-polarized and (back) spinflip states (Fig. 2). Their unperturbed energies we calculate in the half-life calculations according to KLAPDOR et al. (1981). The Hamiltonian to be diagonalized then is

$$H = \frac{G_0}{A} \Sigma\tau\tau \pm \Sigma\Delta\varepsilon_{LS} + \Sigma\Delta_{\text{pair}} + \frac{G_1}{A} \Sigma(\sigma\sigma)(\tau\tau), \tag{1}$$

with

$$\Delta_{\text{pair}} = \langle jj| V |j'j'\rangle_{J=0} = \sqrt{2j+1}\sqrt{2j'+1}\, G \tag{2}$$

where $G = 20/A$ (BERTSCH (1972)), and with (BOHR and MOTTELSON (1969))

$$\Delta\varepsilon_{LS} = 10(2l + 1)\, A^{-2/3}\, \text{MeV}. \tag{3}$$

The matrix elements for the charge plus spin exchange interaction are, in the schematic model (BROWN (1967)),

$$\langle j: \text{CPS, or SFS, or BSFS}|\ G_1\Sigma(\sigma\sigma)\ (\tau\tau)\ |k: \text{CPS, or SFS, or BSFS}\rangle$$
$$= G_1 U^j{}_{\text{CPSorSFSorBSFS}} \cdot U^k{}_{\text{CPSorSFSorBSFS}}/(2J_f + 1), \tag{4}$$

where U^k denote reduced GT transition matrix elements to the unperturbed final states. The reduced transition probability for GT decay to the lth level in the daughter nucleus then is

$$B_l = \frac{1}{2I_i + 1} \left|\sum_k \alpha_l{}^k U^k\right|^2 \tag{5}$$

where $\alpha_l{}^k$ are amplitudes of the wave functions of the daughter states.

Deformation is taken into account by calculating the wave functions of the parent states in a deformed Nilsson potential using a mass dependence of v_{ll}, v_{ls} different for protons and neutrons, according to RAGNARSSON (1970, 1980). The deformation was chosen according to general systematics. The value of G_0 in eqn. (1) is taken from the observed splitting between IAS and AIAS ($G_0 = V_1/4 \approx 30$ MeV). G_1 is extracted from the observed location of GT giant resonances, which is very sensitive to the choice of G_1 in the calculations, in contrast to the low-lying states. A value of $G_1 = 0.8G_0$ is used in our calculations. The renormalization of the strength in the nuclear region of excitation by Δ excitations is not taken into account explicitly for the half life calculations (see, however, section 3.2.3.) but by a subsequent reduction of the strength. This procedure can be justified since the energies of the states carrying the GT strength in the "nuclear" region are only little affected by the coupling to the Δ-hole excitations and since we know the energy dependence of the quenching (GROTZ, KLAPDOR, METZINGER (1983b)).

It should be noted that the effect of the g.s. correlations which we ignored in the β^--decay half life calculations and which are found already negligible by RANDRUP (1973) will further decrease with increasing neutron excess. The occurence of the low-lying back-spin-flip states at large neutron excess justifies the neglection of the contribution of first-forbidden transitions to the half-lives far from stability. It may further be noted that an ansatz using a δ function interaction and keeping both, the direct and exchange term of the ph-interaction (as done by BERTSCH 1980)), when setting the radial matrix elements equal leads to the same results as obtained with the long range force under neglection of the exchange term (ODA (1981)). The above values of G_0 and G_1 are in good agreement with the quantities $\varkappa_\tau$ and $\varkappa_{\sigma\tau}$ used by GAARDE et al. (1981) ($\varkappa_i$ is according to their definition $\varkappa_i = G_i/A$).

The use of a beta strength function $S_\beta(E)$ becomes convenient when large Q_β values will allow the decay to regions of high level densities, i.e. for nuclei far from the stability line:

$$S_\beta{}^{n\lambda}(E) = \frac{1}{\Delta E} \sum_{(i,J,\Delta E)} B_{n\lambda J}(E_i)/D = \sum_J \varrho_J(E)\ \overline{B_{n\lambda J}(E)}/D\ \text{MeV}^{-1}\text{s}^{-1}. \tag{6}$$

$\varrho_J(E)$ is the level density at the excitation energy E in the daughter nucleus, $\overline{B_{n\lambda}(E)}$ is the average reduced transition probability per level around energy E. $S_\beta{}^{-1}(E)$ thus is an ft value per MeV excitation energy.

The beta decay half-life is connected to $S_\beta(E)$ by

$$T_{1/2}^{-1} = \sum_n \int_0^{Q_\beta} S_\beta{}^n(E)\ f_n(Z, Q_\beta - E)\ \mathrm{d}E \tag{7}$$

and for the branching ratio $P_{\beta DP}$ for a β-delayed decay process, like β-delayed particle emission or fission, assuming that this occurs after full equilibration of the emitter nucleus

$$P_{\beta DP} = T_{1/2} \sum_{n} \sum_{J} \int_{0}^{Q_\beta} S_\beta^{\pi, J}(E)\, f_n(Z, Q_\beta - E)\, W_P^{J,\pi}(E)\, dE, \tag{8}$$

where $W_P^{J,\pi}(E)$ is the probability for particle emission or fission from a state at energy E with spin J^π.

2.2. Results

2.2.1. Beta strength functions

As examples the GT strength functions calculated (Grotz et al. (1983)) for the decay of ^{76}Ge, ^{128}Te and ^{208}Pb are shown in Fig. 3. Calculations for ^{128,130}Te and also for ^{76}Ge are of interest in the discussion of double beta decay (see section 3.2.3). From comparison of the ^{208}Pb strength function with a (p, n) experiment it has been realized for the first time that the shape of the beta strength function can be sensitive to the nuclear charge distribution (see section 3.2.4).

2.2.2. Beta decay half-lives

The calculated beta strength functions were used to calculate the beta decay half-lives (eqn. 7) for nuclei with $Z = 6-114$ between beta stability line and neutron drip line. It should be noted that the results are — as those for $S_\beta(E)$ — independent of nuclear mass formulae when determining the unperturbed energies of the basis states in the daughter nucleus as mentioned above. The calculations reproduce rather well the experimental data. Figures 4—6 show some examples. The degree of agreement with the experimentally known half-lives gives a feeling for the predictive power of the calculations. Defining $\bar{x} = N^{-1}\Sigma_{i=1} x_i$ with x_i denoting the ratio of experimental to calculated half-life for $T_{1/2}$ (exp) $\geqq T_{1/2}$ (calc), and its inverse for $T_{1/2}$ (exp) $< T_{1/2}$ (calc), the

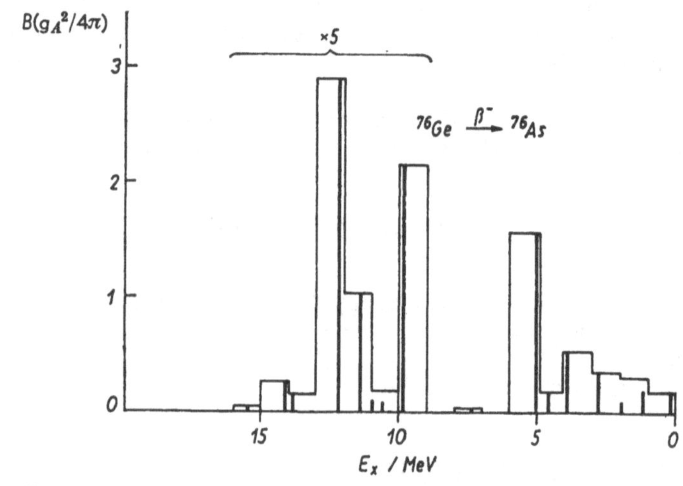

Fig. 3a

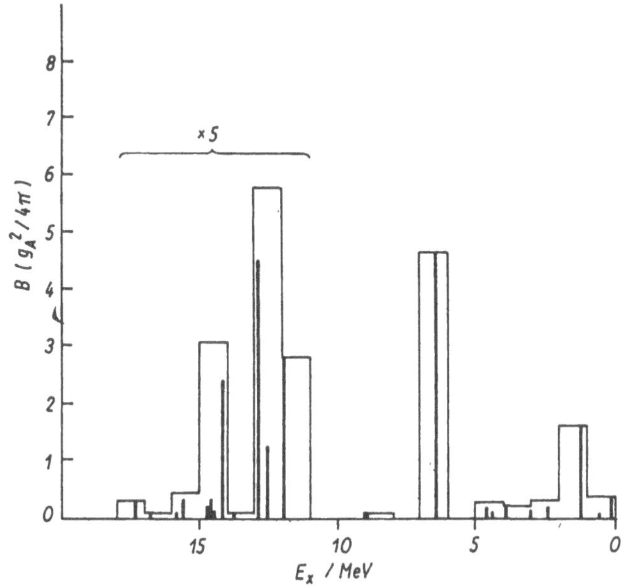

Fig. 3b $^{128}Te \xrightarrow{\beta^-} {}^{128}I$

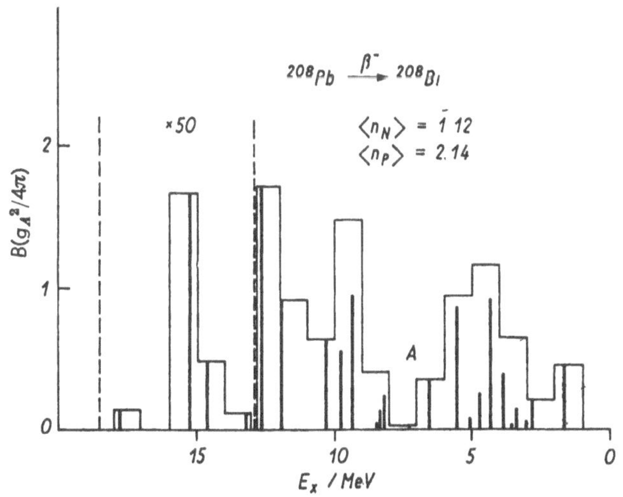

Fig. 3c

Fig. 3. Calculated GT strength functions (including pairing and spin-isospin correlations in the parent nucleus up to four quasi-particle admixtures) for beta decay of (a) ^{76}Ge (b) ^{128}Te (c) ^{208}Pb. The first two cases are of interest for double beta decay (see section 3.2.3.). The energies are relative to the g.s. of the daughter nuclei, The strength has to be multiplied by the given factors, as indicated (for ^{208}Pb within the dashed lines).

situation is the following. From the known nuclei undergoing β^- decay with $T_{1/2} \leqq 10^6$ s (502 cases) 79% are reproduced with an average deviation of $\bar{x} = 2.45$, from these 47% with $x_i < 2$ and $\bar{x} = 1.36$. From the known nuclei with $T_{1/2} \leqq 60$ s (262 cases) 95% are reproduced with $\bar{x} = 1.96$, from these 65% with $x_i < 2$ and $\bar{x} = 1.36$. The accuracy of the calculation thus increases with increasing distance from the beta stability line. The

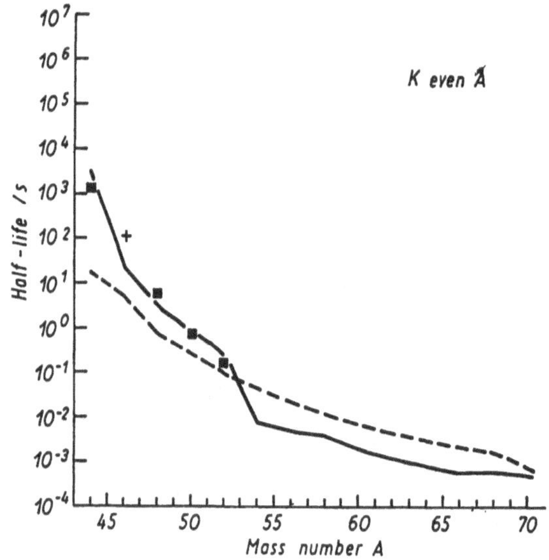

Fig. 4a

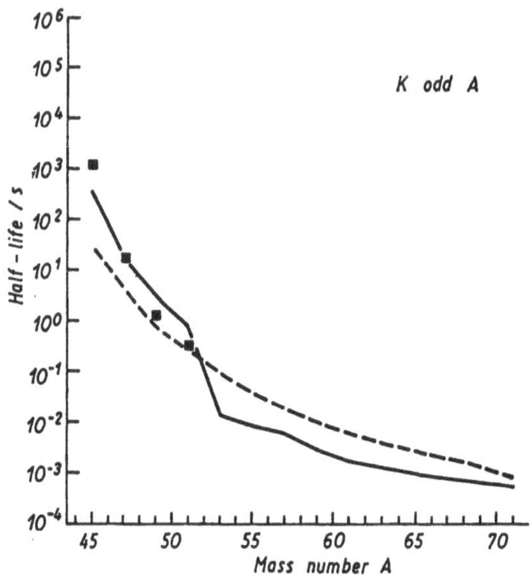

Fig. 4b

Fig. 4. Calculated (solid lines) and measured (squares) beta decay half-lives of K isotopes up to the neutron drip line (from KLAPDOR, et al. (1981c, 1982)). The dashed lines correspond to the 'gross theory'

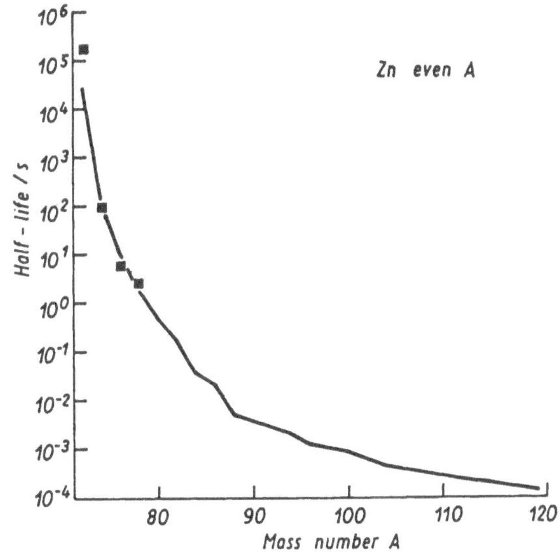
Zn even A
Half - life / s
10^6
10^5
10^4
10^3
10^2
10^1
10^0
10^-1
10^-2
10^-3
10^-4
80
90
100
110
120
Mass number A

Fig. 5a

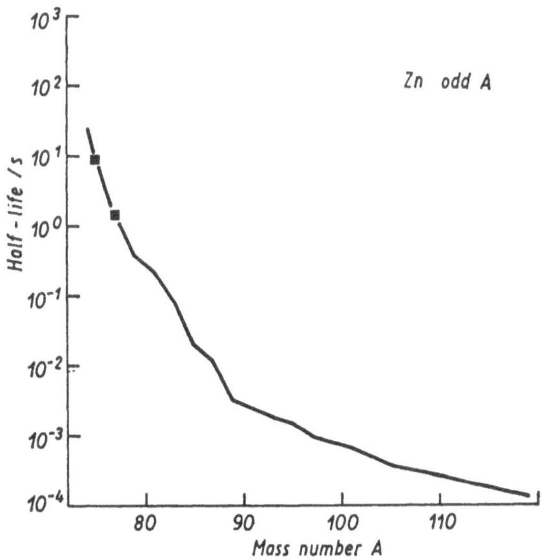
Zn odd A
Half - life / s
10^3
10^2
10^1
10^0
10^-1
10^-2
10^-3
10^-4
80
90
100
110
Mass number A

Fig. 5b

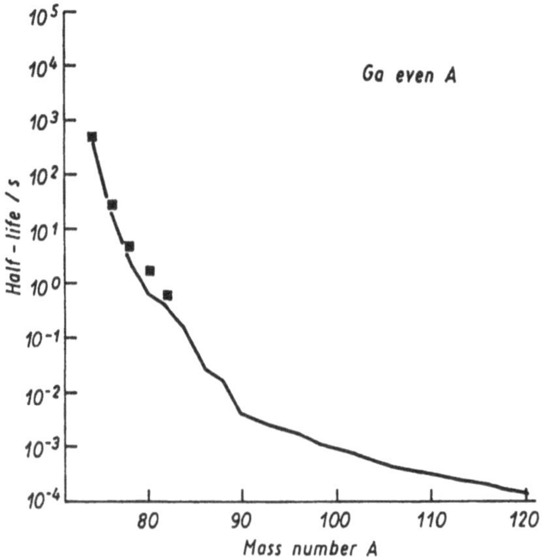
Ga even A
Half-life / s
10^5
10^4
10^3
10^2
10^1
10^0
10^{-1}
10^{-2}
10^{-3}
10^{-4}
80
90
100
110
120
Mass number A

Fig. 5c

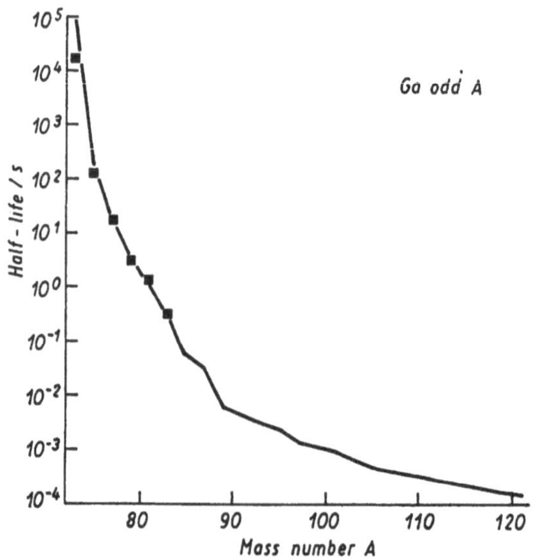
Ga odd A
Half-life / s
10^5
10^4
10^3
10^2
10^1
10^0
10^{-1}
10^{-2}
10^{-3}
10^{-4}
80
90
100
110
120
Mass number A

Fig. 5d

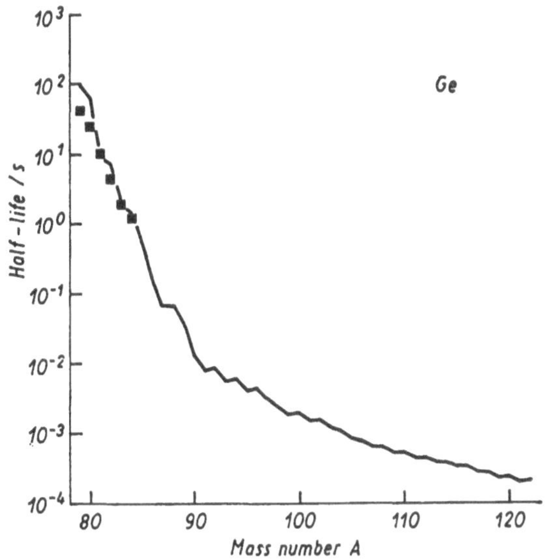

Fig. 5e

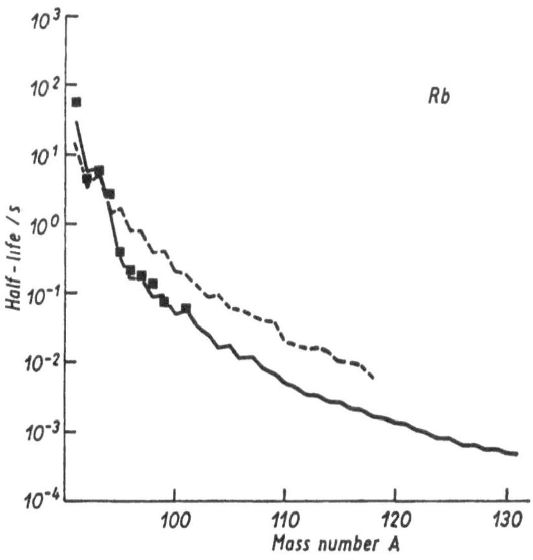

Fig. 5f

Fig. 5. Calculated (solid lines) and measured (squares) beta decay half-lives of isotopes of Zn, Ga, Ge and Rb (from KLAPDOR et al. (1981c, 1982)). The dashed line corresponds to the gross theory

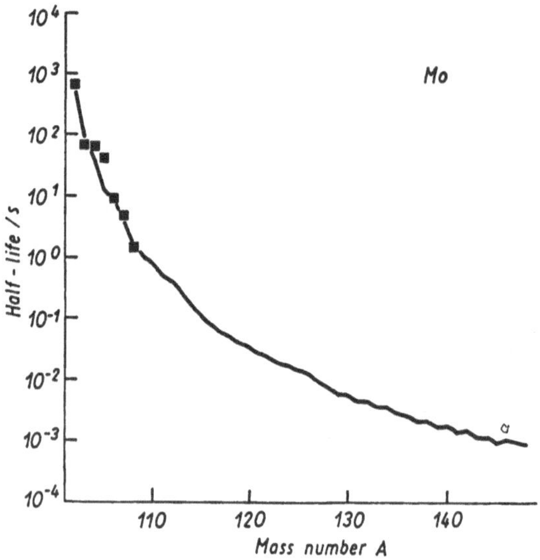
Mo
Half - life / s
10^4
10^3
10^2
10^1
10^0
10^-1
10^-2
10^-3
10^-4
110
120
130
140
Mass number A

Fig. 6a

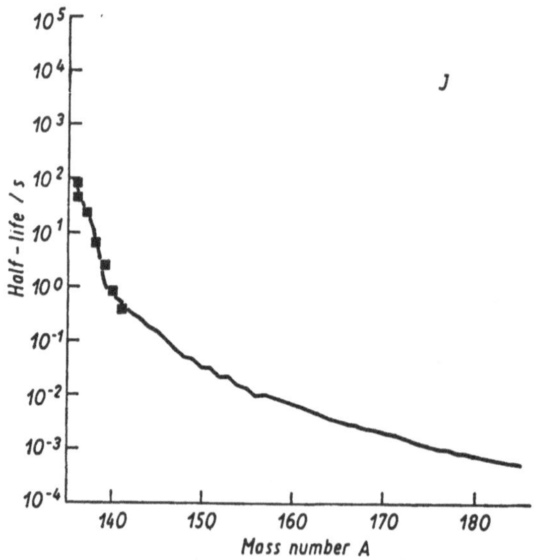
J
Half - life / s
10^5
10^4
10^3
10^2
10^1
10^0
10^-1
10^-2
10^-3
10^-4
140
150
160
170
180
Mass number A

Fig. 6b

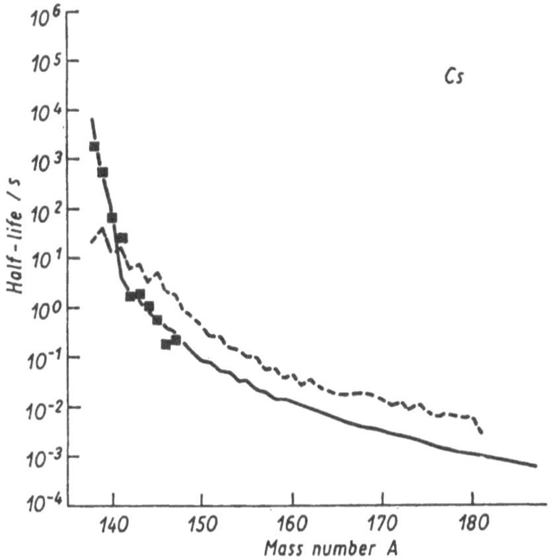
Cs
Half-life / s
10^6
10^5
10^4
10^3
10^2
10^1
10^0
10^{-1}
10^{-2}
10^{-3}
10^{-4}
140
150
160
170
180
Mass number A

Fig. 6c

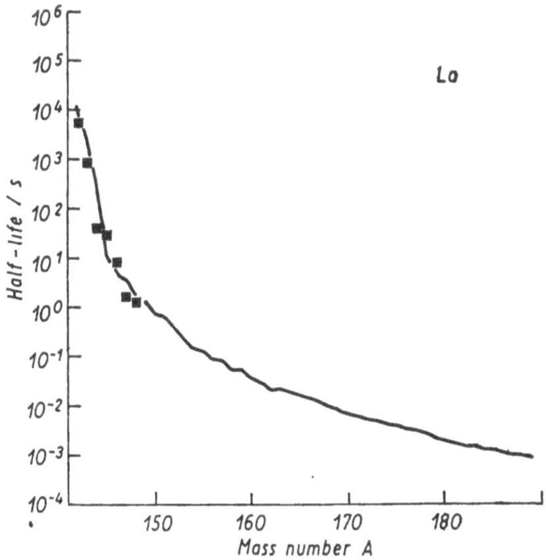
La
Half-life / s
10^6
10^5
10^4
10^3
10^2
10^1
10^0
10^{-1}
10^{-2}
10^{-3}
10^{-4}
150
160
170
180
Mass number A

Fig. 6d

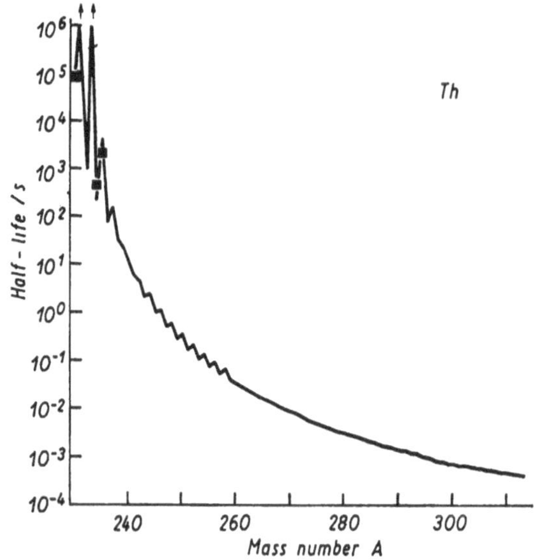

Fig. 6e

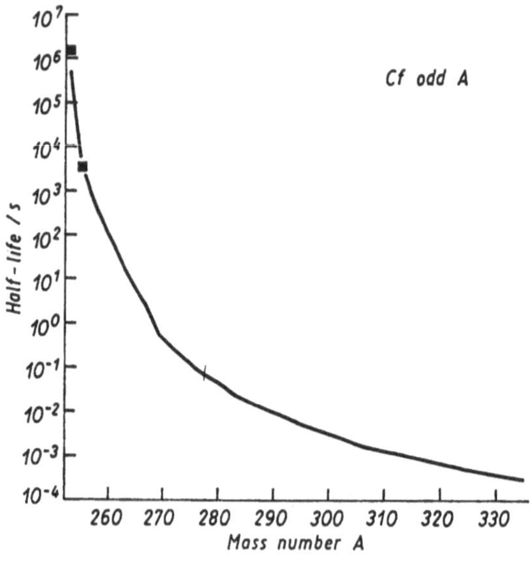

Fig. 6f

Fig. 6. Calculated (solid lines) and measured (squares) beta decay half-lives of neutron-rich isotopes of Mo, J, Cs, La, Th and Cf (from KLAPDOR et al. (1981 c, 1982)). The dashed line corresponds to the gross theory

reason is that for lower Q_β values the sensitivity of the half-life to the calculated beta strength distribution increases very strongly. It is thus the latter number ($\bar{x} \approx 1.3$) which is representative for the accuracy of the predictions to the nuclei with unknown half-lives. Recent calculations by ALEKSANKIN et al. (1981) and GAPONOW and LYUTOSTANSKII (1981) reproduce the half-lives of some 50 selected nuclei with $T_{1/2} \lesssim 10$ s on the average within 50—80%. The microscopically calculated half-lives are systematically

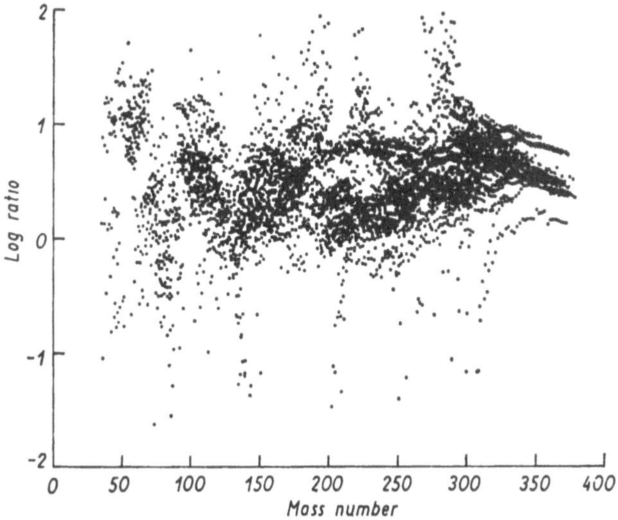

Fig. 7. Ratio of beta decay rates (as function of mass number A) calculated by KLAPDOR et al. (1981c, 1982) and by the gross theory (this figure was prepared by A. G. W. CAMERON)

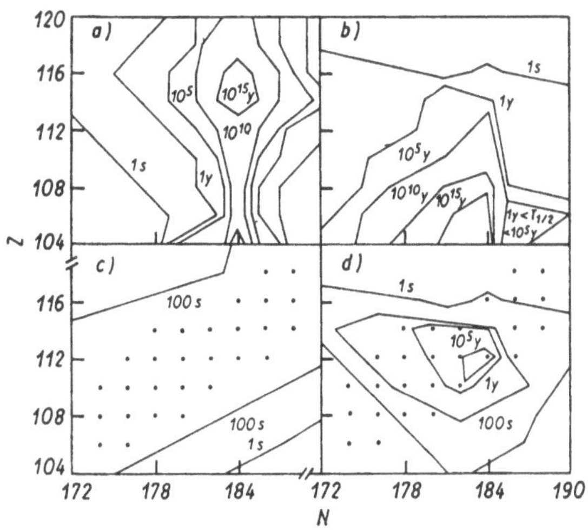

Fig. 8. Contour plots of calculated half-lives for spontaneous fission (a) α decay (b) β decay (c) and the total half-lives (d) of even superheavy nuclei. (a), (b) and the half-lives for neutron-deficient nuclei in (c), (d) are from FISET and NIX (1972), the β^- decay half-lives in (c) and the total half-lives of neutron-rich nuclei in (d) are from KLAPDOR, et al. (1981c, 1982)

shorter, up to an order of magnitude and more, for neutron-rich nuclei than those predicted by the gross theory of TAKAHASHI et al. (1973), which on the other hand tends to underestimate the half-lives of nuclei close to the beta stability line. The difference between the predictions of the two theories depends in detail on Z and A. Fig. 7 shows for $A \sim 30-250$ the ratios of our calculated beta decay rates to those of the gross theory. It might be noted that the half-lives we obtain for the nuclides around the island of superheavy elements are also systematically shorter than those calculated, still with the assumption $S_\beta(E) \sim \varrho(E)$, by Fiset and NIX (1972) (see Fig. 8). With their half-lives for α-decay and spontaneous fission we find (Fig. 8) as longest-lived nucleus of the superheavy island $Z = 112$, $N = 184$ with $T_{1/2}^{\text{total}} = 10^{5.4}$ a, instead of $Z = 110$, $N = 182$ with $T_{1/2}^{\text{total}} = 10^{9.4}$ a given by FISET and NIX (1972). The whole body of our β^- decay half-lives is available as Internal Report (KLAPDOR et al. (1981c)) and has been recently published (KLAPDOR, METZINGER, ODA (1984)).

2.2.3. Beta-delayed neutron emission and fission rates

The branching ratio for beta-delayed neutron emission and fission can be written

$$P_n = T_{1/2} \cdot \int\limits_{B_n}^{Q_\beta} \frac{T_n(E)}{T_n(E) + T_f(E) + T_\gamma(E)} S_\beta(E)\, f(Q_\beta - E)\, \mathrm{d}E \tag{9}$$

$$P_{\beta df} = T_{1/2} \cdot \int\limits_{0}^{Q_\beta} \frac{T_f(E)}{T_n(E) + T_f(E) + T_\gamma(E)} S_\beta(E)\, f(Q_\beta - E)\, \mathrm{d}E\,. \tag{10}$$

where T denote total transmission coefficients.

For the calculation of the neutron transmission coefficients we used an optical model potential obtained (JEUKENNE et al. (1977)) from infinite nuclear matter calculations in the Brückner-Hartree-Fock approximation with a Reid's hard core nucleon-nucleon interaction by applying the local density approximation to finite nuclei. Such a "next to first principles" calculation should allow a more reliable extraplation into the region of unstable nuclei than phenomenological models fitted to stable nuclei (see THIELEMANN et al. (1983)).

The transmission coefficients for fission were calculated by the method described by BACK et al. (1974), LYNN (1978), i.e. by assuming a "complete damping mechanism" which means averaging over transmission resonances assuming equal level spacing in the second minimum:

$$\begin{aligned} P_f(E, J, \pi) &= \frac{T_f(E, J, \pi)}{T_{\text{tot}}(E, J, \pi)} \\ &= \left\{1 + \left(\frac{T_n + T_\gamma}{T_{AB}}\right)^2 + 2\left(\frac{T_n + T_\gamma}{T_{AB}}\right) \coth\left|\frac{1}{2}(T_A + T_B)\right|\right\}^{-1/2} \end{aligned} \tag{11}$$

where

$$T_{AB}(E, J, \pi) = \frac{T_A(E, J, \pi)\, T_B(E, J, \pi)}{T_A(E, J, \pi) + T_B(E, J, \pi)} \tag{12}$$

$$T_A = \int\limits_0^E T_{HW}\left(E - E_A - \frac{\hbar^2}{2\theta} l(l+1) - \varepsilon, \hbar\omega_A\right) \varrho_A(E, J, \pi)\, \mathrm{d}\varepsilon \tag{13}$$

and

$$T_{HW}(E, \hbar\omega) \mp \left(1 + \exp\left(-2\pi E/\hbar\omega\right)\right)^{-1}. \tag{14}$$

The energies of the axial asymmetric and mass-asymmetric saddle-points E_A and E_B were taken from HOWARD and MØLLER (1980), the values of $\hbar\omega_A$ and $\hbar\omega_B$ according to BACK et al. (1974), LYNN (1978) ($\hbar\omega_A = 1.00$, 0.80, 0.60 and $\hbar\omega_B = 0.60$, 0.52 and 0.42, for even-even, odd mass, and odd-odd nuclei, respectively). The level densities beyond the saddle points were calculated by the back-shifted Fermi gas formula modified, following BRITT (1980) and LYNN (1978), by empirical factors, which take care of the increase at the saddle points due to the increased deformation and the resulting coupling to low-lying rotational excitations: $\varrho_A \approx 5\varrho$, $\varrho_B \approx 2\varrho$; where ϱ denotes the level density for a deformation corresponding to that of the g.s.

The calculations of P_n, P_{1n}, P_{2n} and $P_{\beta df}$ have been performed (THIELEMANN et al. (1983)) over the whole nuclidic chart. The mass formula of HILF et al. (1976) has been used which has proven to show the best overall agreement with experimental masses far from stability. Its predictions are very close to the new formula by MØLLER and NIX (1981) which is suited particularly for nuclei far from stability; unfortunately the compilation by these authors does not give all the masses needed for an r-process calculation.

Table 1

Calculated and measured P_n and P_{2n} values for neutron-rich Na, Rb and Cs isotopes

Nucleus	P_n (experiment)	P_n (calc.)	P_{2n}/P_n (exp.)	P_{2n}/P_n (calc.)
^{30}Na	29 ± 4†	11.6	0.042 ± 0.008†	0.04
^{31}Na	30 ± 8†	33.2	0.023 ± 0.005†	0.005
^{32}Na	13 ± 4†	23.3	0.24 ± 0.05†	0.11
^{92}Rb	0.011 ± 0.001*	0.009		
^{93}Rb	1.40 ± 0.08*	1.57		
^{94}Rb	10.1 ± 0.6*	9.24		
^{95}Rb	8.9 ± 0.6*	11.47		
^{96}Rb	13.5 ± 0.9*	23.44	$< 2.8 \cdot 10^{-4}$††	$3.7 \cdot 10^{-4}$
^{97}Rb	25.2 ± 1.8**	21.57	$< 3.2 \cdot 10^{-4}$††	$1.6 \cdot 10^{-5}$
	21.5 ± 2.5†††			
^{98}Rb	18.4 ± 2.9**	23.91	$(3.8 \pm 0.6)\ 10^{-3}$††	$4.8 \cdot 10^{-3}$
^{141}Cs	0.029 ± 0.002*	0.014		
^{142}Cs	0.097 ± 0.007*	0.07		
^{143}Cs	1.74 ± 0.12**	2.04		
^{144}Cs	2.95 ± 0.25**	14.08		
^{145}Cs	13.6 ± 1.0*	23.67		
	21.8 ± 1.5**			

† Detraz, et al. (1980); * LUND, et al. (1980); ** RISTORI, et al. (1979); *** Reeder, et al. (1977); †† Reeder, et al. (1981); ††† Engler, and NEEMANN (1981).

In Table 1 calculated and experimental P_n and P_{2n} values are compared for Na, Rb and Cs isotopes. The agreement is generally somewhat weaker than found for the half-lives since the P_n values are (by the neutron separation energy) dependent on nuclear masses. Fig. 9 shows in an Z,N-plane contour lines for $P_n = 0.5$, and for $P_{2n} = 0.2 P_{1n}$. Half-way between the curves for $P_n = 0.5$ and $P_{2n} = 0.2 P_{1n}$ almost 100% neutron emission ($P_n = 1$) is reached. Comparison with the fission yields of Fig. 11 shows that close to the neutron drip line beta-delayed neutron emission dominates again over β-delayed fission.

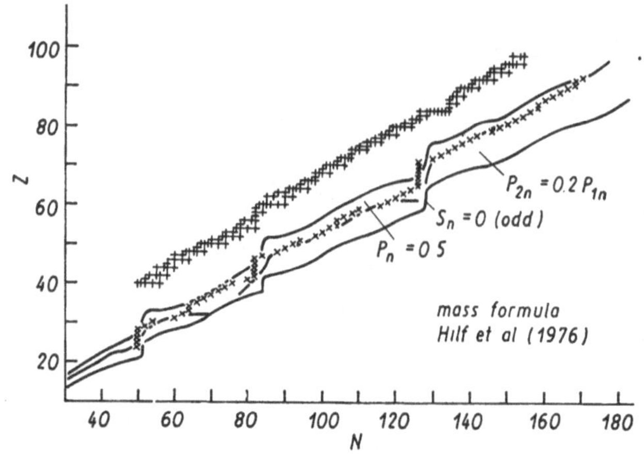

Fig. 9. Contour lines of beta-delayed neutron emission rates $P_n = 0.5$ and $P_{2n} = 0.2P_{1n}$ calculated with microscopic $S_\beta(E)$ · $S_n = 0$ denotes the neutron drip line (for odd A), ×: the r-process path in explosive He-burning (see section 3.1), +: the line of β-stability (from THIELEMANN et al. (1983))

Fig. 10 shows the situation for one of the few cases of β^- delayed fission for which experimental data exist (GANGRSKY et al. (1978)), namely ^{236,238}Pa (see, however, KLAPDOR et al. (1979) and section 3.2.1 for β^+ delayed fission). The calculations were done using the experimental barriers (BRITT (1980), BACK et al. (1974)). The calculated values of $P_{\beta df} = 5 \cdot 10^{-9}$ and $6 \cdot 10^{-7}$ for ^{236}Pa and ^{238}Pa, respectively, agree, within the experimental uncertainties of the barriers, with the measured (GANGRSKY et al. 1978) values $P_{\beta df} = 10^{-9} - 3 \cdot 10^{-10}$ and $6 \cdot 10^{-7}$—10^{-8}, respectively. The tendency of slightly too large theoretical values is understandable from the fact, that the approach treating fission as a tunneling through separate barriers becomes marginal at least for ^{236}Pa,

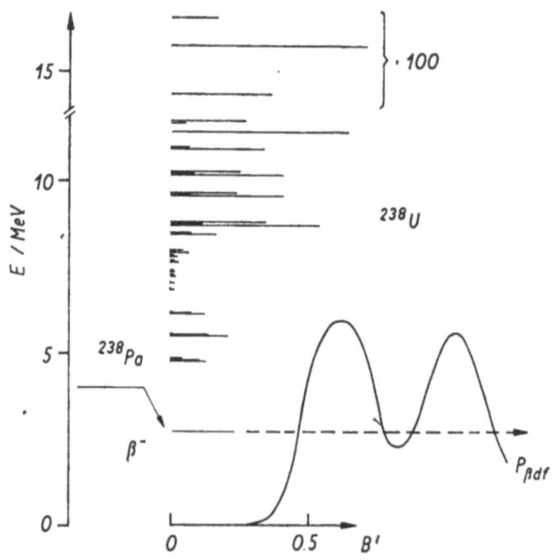

Fig. 10. The calculated beta strength function for the decay of ^{238}Pa used in the calculation of the beta-delayed fission yield of ^{238}U

leading to an overestimate of the fission probability. This problem does not exist for large $P_{\beta df}$ values.

Fig. 11 gives calculated ratios of β-delayed fission $\geqq 1\%$ over the nuclidic chart between the β stability line and the neutron drip line. The uncertainties resulting from the fission barriers (see HOWARD and MØLLER (1980)) were estimated by changing the heights of the first and second barrier, respectively, and can range up to a factor of 5 for small $P_{\beta df}$ values ($\leqq 5\%$) but becomes relatively small for larger values (THIELEMANN et al.

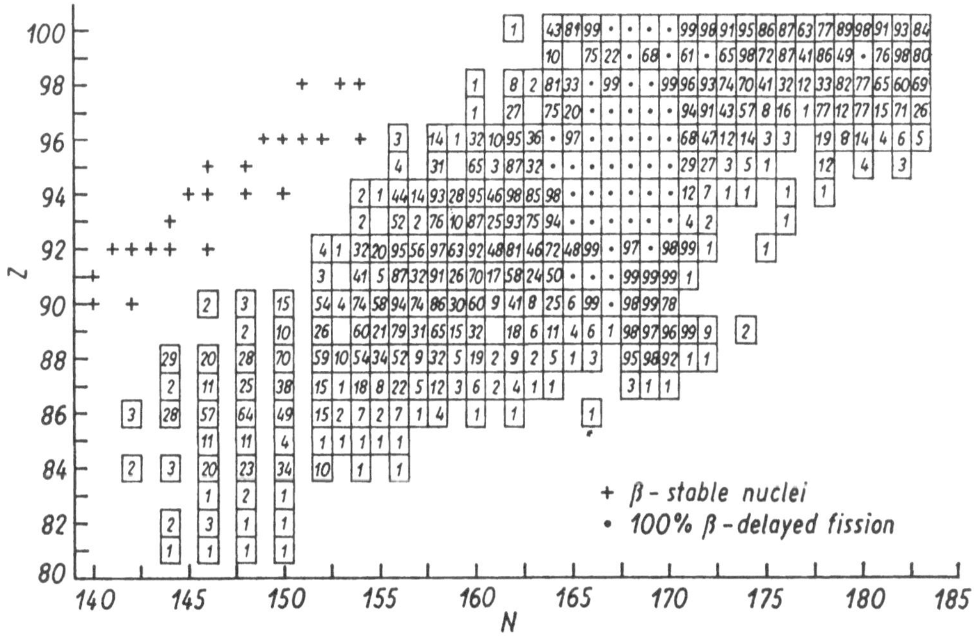

Fig. 11. (a) The range of the nuclidic chart (for $Z \leqq 100$) where the β^- delayed fission rate $P_{\beta df}$ is $\geqq 1\%$. The calculations were performed with the microscopic $S_\beta(E)$ from KLAPDOR et al. (1981c, 1982), the fission barrier heights of HOWARD and MØLLER (1980) and the mass formula of HILF et al. (1976). The daughter nuclei reached in β^- decay are labelled by the $P_{\beta df}$ value; dots denote $P_{\beta df} = 100\%$, crosses denote the beta stability line. An island of 100% beta-delayed fission is predicted around $Z = 94$ and $N = 168$ (from THIELEMANN et al. (1983))

(1983)). In particular, the island of 100% beta-delayed fission around $Z = 94$, $A = 168$ (which occurs as a result of the second minimum lying lower than the first minimum so that only tunnelling through the second barrier is needed) remains rather stable. This is important for some of the consequences to be discussed in section 3. This island does not exist in the schematic calculations of WENE and JOHANSSON (1976) who described the double-humped fission barrier by one barrier and still used an $S_\beta \sim \varrho(E)$. KODAMA and TAKAHASHI (1975), who used the gross theory of β-decay, obtained by far smaller $P_{\beta df}$ values than given in Fig. 11.

The island of 100% of beta-delayed fission persists also when different mass formulae are used in the calculation of $P_{\beta df}$ values (see THIELEMANN et al. (1983), KLAPDOR (1983).

3. Consequences

With the results of Section 2 there seems to exist for the first time a reliable set of beta-decay half-lives and rates of β-delayed neutron emission and fission for neutron-rich nuclei. What are the consequences? This will be discussed in the following subsections for some problems of astrophysics and nuclear physics.

3.1. Astrophysics

There are several processes in astrophysics, which are sensitively dependent on beta decay properties. We shall discuss three of them:

(1) The astrophysical *r*-process (rapid neutron capture process). It is generally accepted that this process, together with the *s*- and *p*-processes (BURBIDGE et al. (1957), CAMERON (1957), for a review see TRIMBLE (1975, 1982, 1983)) is responsible for the synthesis of the heavy elements in the universe. The question of the astrophysical site in the universe of the *r*-process was, however, an open problem in astrophysics for more than two decades (see e.g. OHNISHI (1979), NORMAN and SCHRAMM (1979)). It will be shown below that the new β-decay half-lives allow for the first time to give a solution of this longstanding problem.

(2) Cosmochronology by the so-called chronometric pairs. This method which in principle allows to determine the age of the universe and details of the galactic and solar system evolution, can be put for the first time on a reliable basis as consequence of the solution found for the *r*-process and by including the effect of beta-delayed fission.

(3) Detection of solar and galactic neutrinos. The efficiency as function of neutrino energy of the detectors used or to be used in the future (e.g. in the Gallium solar neutrino experiment) is determined by the beta strength function.

3.1.1. The astrophysical *r*-process

The high neutron densities needed to drive an *r*-process can be produced only in explosive events. Consequently, in searching for its astrophysical site dynamical calculations are necessary. This means that the results are sensitive to the beta decay rates of neutron-rich nuclei. The range of nuclei for which the β-decay properties are needed extends from the beta stability line almost out to the neutron drip line as illustrated schematically in Fig. 12. Beta-delayed neutron emission contributes to a smoothing of the *r*-process yield curve of stable elements (KODAMA and TAKAHASHI (1973), BLAKE and SCHRAMM (1973)). Beta-delayed fission influences the production rates of heavy elements, particularly the production ratios of chronometric pairs (KLAPDOR and WENE (1979)).

Up to now, mainly two sites for the *r*-process have been discussed in the literature, which could produce the necessary neutrons, both being associated with supernova explosions (see Fig. 13). In view of the high neutron densities required, BURBIDGE et al. (1957) and SEEGER et al. (1965) suggested the supernova core as the site of the *r*-process. The first test of this "classical" *r*-process site with realistic supernova-models was performed by HILLEBRANDT et al. (1976, 1978). Their hydrodynamical explosion calculations for the outer part of the core yielded, for a special choice of the radial dependence of the neutron to proton ratio and of the position of the mass-cut between the remaining neutron star and the ejected envelope, *relative* solar *r*-abundances in one event. However, the neutron to proton ratio was a rather unknown quantity at that time. Further, the mass-cut which had to be assumed for reproducing the solar *r*-abundances, led to a tremendous overproduction (by a factor of 10^3-10^4) of r-nuclei in this scenario (HILLEBRANDT (1978, 1979)). Taking into account the effect of weak neutral currents (see, e.g. TUBBS and SCHRAMM (1975), FREEDMAN (1974)) reduces, however, the n/p ratio to such an extent that this "classical" r-process scenario is practically ruled out (HILLEBRANDT (1982)).

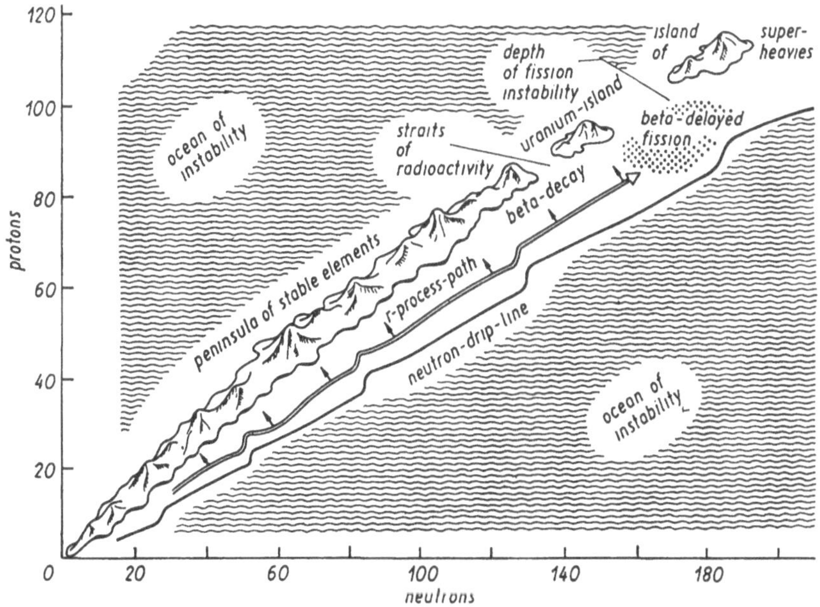

Fig. 12. The range of nuclei whose β-decay properties are needed for calculations of the synthesis of elements in the universe by rapid neutron capture (r-process) extends from the line of beta stability almost out to the neutron drip line: The element distribution built up for less than a second in supernova explosions is given by the r-process path. This distribution decays back by β-decay to the β stability line when the high neutron density built up in the explosion drops

The fact that the "classical" r-process site cannot reproduce the cosmic r-abundances under physically plausible conditions, has led to the proposal of the r-process by *explosive Helium burning* (HILLEBRANDT and THIELEMANN (1977), TRURAN et al. (1978)): During hydrostatic helium burning the ^{14}N left from hydrogen-(CNO) burning is transformed into ^{22}Ne via the reaction chain $^{14}N(\alpha, \gamma)^{18}F(\beta^+)^{18}O(\alpha, \gamma)^{22}Ne$. When the supernova detonation shock front passes the He-burning shell, the reaction $^{22}Ne(\alpha, n)^{25}Mg$ (which is driving the s-process during hydrostatic helium burning) produces for a short time a neutron density, which drives an r-process. That this process proceeds far from the center of the supernova core makes its investigation conveniently independent of still unsolved problems (see HILLEBRANDT (1981), LATTIMER (1981)) of the triggering of the supernova explosion by the gravitational collapse.

In this scenario we obtain (THIELEMANN et al. (1983)) the result given in Fig. 14. Shown is the calculated abundance by mass, in comparison to the observed solar system r-element distribution. Obviously a solar r-distribution can be obtained only with the microscopic β-decay half-lives. The solar r-distribution is built up within 0.29 s. The gross theory half-lives would request a longer neutron exposure which is excluded in this scenario. Fig. 15 shows the result after allowing for further decay of the nuclei in the beta stability line by α-decay and spontaneous fission over 10^9 and 10^{10} a. From the trans-bismuth nuclei only the "cosmochronometers" are still present (see next section). It

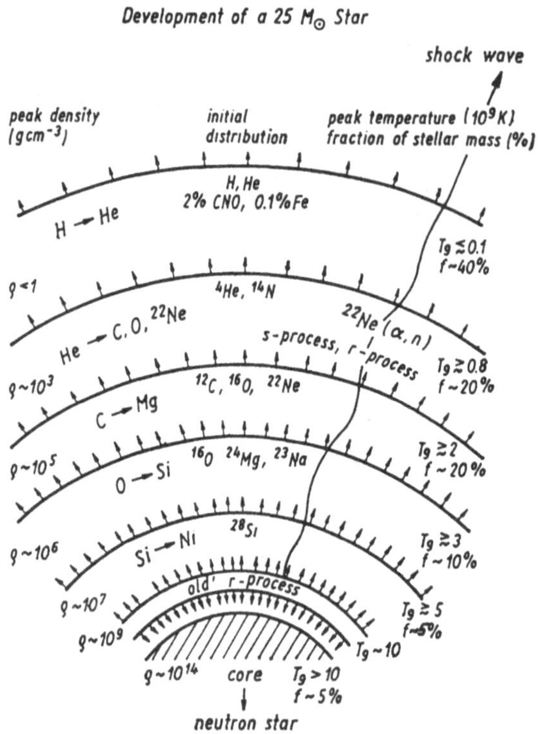

Fig. 13. Schematic representation of the structure and development of a heavy star of $\sim$25 solar masses). Hydrostatic burning has led to the typical onion shell structure and built up elements up to Fe, Ni. Gravitational collaps of the core leads to formation of a neutron star and ejection of $\sim$95% of the stellar mass (supernova explosion). The ejected outer shells are passed by a detonation shock front which induces explosive shell burning. The latter modifies the preshock element distributions (for the inner shells this is discussed in detail by WOOSLEY and WEAVER (1981)). In the He-burning shell in which during hydrostatic burning the s-process is proceeding the passing shock front leads to explosive He-burning and an r-process which synthesizes the heavy elements in the universe (KLAPDOR et al. (1981))

may seem from Fig. 15 that the calculated abundances of Pb and Bi do not agree with the solar r-abundances of CAMERON (1973) displayed in the same figure. However, as pointed out by THIELEMANN et al. (1983), in the determination of the r-abundances by subtracting s-abundances from the solar abundance distribution neither CAMERON (1973) nor the recent works by KÄPPELER et al. (1982) and CAMERON (1982) take into account the cycling at the termination of the s-process ($^{209}Bi(n, \gamma)^{210}Bi(\beta^-)^{210}Po(\alpha)^{206}Pb(n, \gamma)$ $^{207}Pb(n, \gamma)^{208}Pb(n, \gamma)^{209}Pb(\beta^-)^{209}Bi\ldots$). When this is done, WARD and CLAYTON (1982) find a flat abundance curve for the stable Pb and Bi isotopes with a slight decrease from ^{206}Pb to ^{208}Pb similar to that predicted in Fig. 15b. Fig. 16 demonstrates the strong effect, beta delayed neutron emission has in the smoothening of the odd-even staggering of the final element distribution and the effect of beta-delayed fission on the final element distribution. Fig. 17 shows the position of the calculated r-process path in the upper part of the (Z,N)-plane. Its cut-off is obviously determined by beta-delayed fission at already relatively small Z (around $Z = 92$). Consequently no superheavy elements will be produced in the r-process in contrast to earlier speculations of SCHRAMM and FOWLER

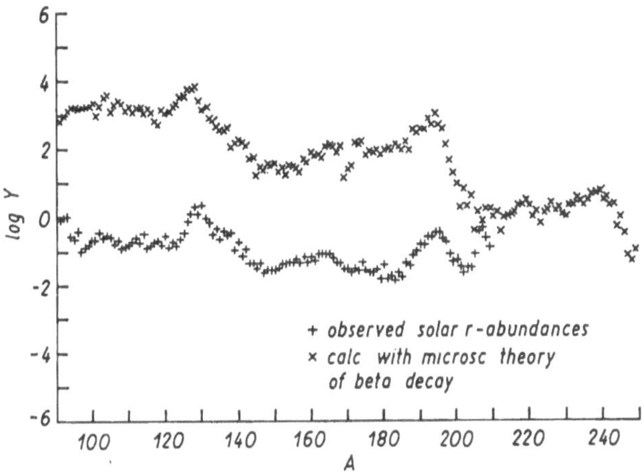

Fig. 14a

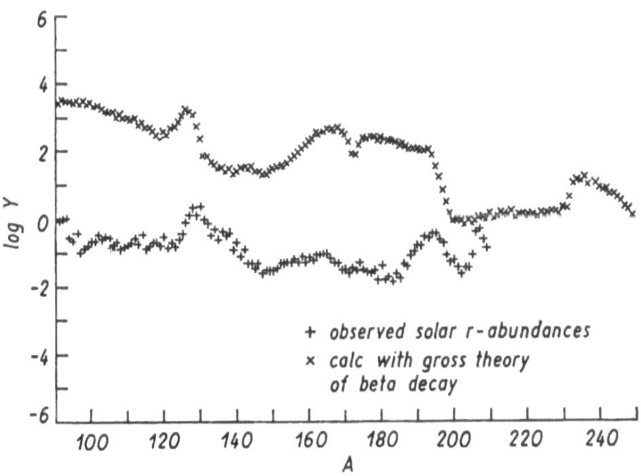

Fig. 14b

Fig. 14. ×: Computed r-process abundance distributions in explosive He-burning at freeze-out, +: observed solar system r-process abundances (the latter from CAMERON (1973)). All abundances are normalized to log Y(Si) = 6. (a) Beta decay half-lives from KLAPDOR et al. (1981c, 1982). Beta-delayed neutron and fission rates from THIELEMANN et al. (1983). Excellent relative agreement is obtained with the observed solar r-element distribution. (b) Beta decay half-lives from the gross theory of β-decay (TAKAHASHI et al. 1973). The solar r-element distribution is not reproduced

(1971) but in support of HOWARD and NIX (1974), BOLEU et al. (1972), SCHRAMM et FISET (1973), whose conclusion based on the short spontaneous fission half-lives during the decay back.

The new site avoids the problem of the "classical" r-process of overproduction of heavy r-elements. Not only the calculated relative r-element abundances agree very well with the observation, also the calculated absolute abundances, which are in (part of) the He shell about a factor of 10^{3-4} larger than in normal matter and correspond to an

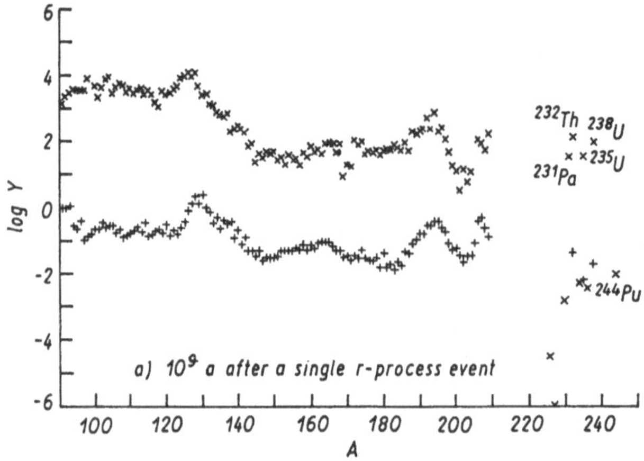

Fig. 15a

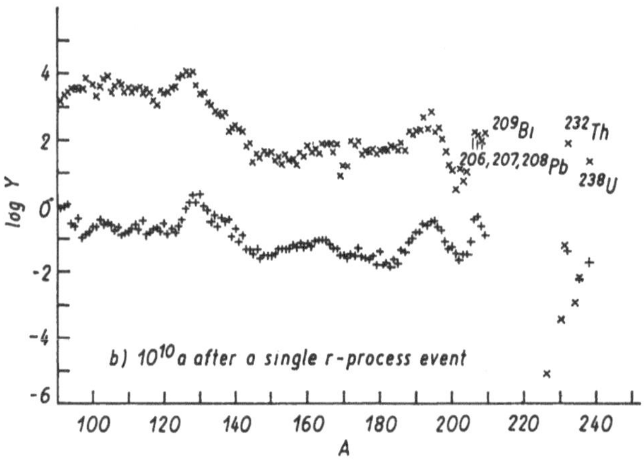

Fig. 15b

Fig. 15. The computed r-process abundance distribution of Fig. 14a after allowing for further decay by α emission and spontaneous fission in the β-stability line for 10^9 (a) and 10^{10} years (b). From the transbismuth nuclei only the cosmochronometers remain (see section 3.1.2.) (from Thielemann et al. 1983)

average enhancement factor of r-material of 10—15 for a $25 M_\odot$ star, are comparable with the enhancement factor predicted (Woosley and Weaver (1981)) for the production of lighter elements in explosive burning of inner shells. The new r-process also facilitates the explanation of observed r-element overabundances in meteorites (Klapdor (1984)).

For comments to some recent papers (Blake et al. (1982), Cowan et al. (1982)) also discussing explosive He shell-burning (but not using microscopic beta-decay half-lives) see Thielemann et al. (1983), Klapdor (1983a).

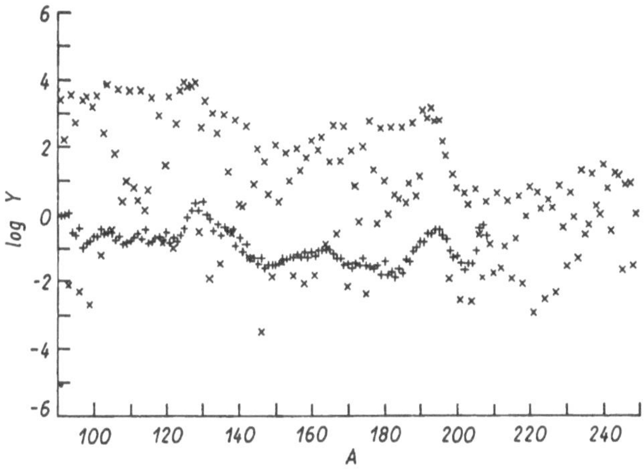

Fig. 16a

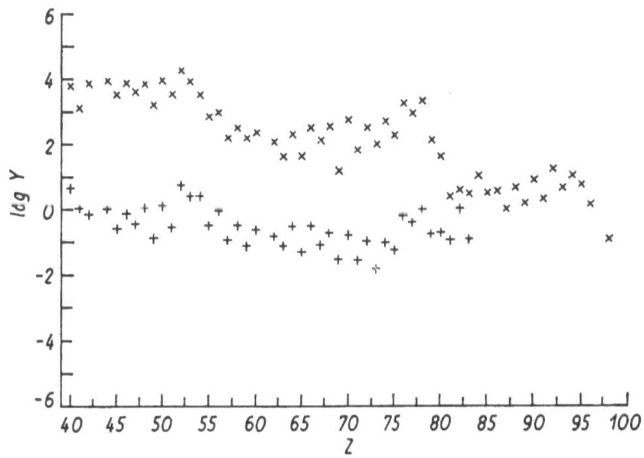

Fig. 16b

Fig. 16. (a) Illustration of the smoothing effect of beta-delayed neutron emission on the r-process yield of stable elements. The same calculation as in Fig. 14a but without β-delayed neutron emission. (b) The effect of beta-delayed fission on the production of beta-stable elements in the r-process. Shown is the same calculation as in Fig. 14a but the abundance versus proton number Z

3.1.2. Cosmochronology

There are three independent methods to determine the age of the galaxy and of the universe: (1) Cosmological models which extrapolate the time of the big bang from the Hubble constant. (2) Determination of the ages of the oldest stars, in particular, the ages of globular clusters. (3) Nucleochronology, that is direct dating of the time of nucleosynthesis of radioactive nuclei from observed or implied abundances. For a recent review see Symbalisty and Schramm (1981). Our results on the r-process and on the beta decay of neutron-rich nuclei allow an improvement of the third of these methods, as qualitatively already discussed by Klapdor (1976), Klapdor and Wene (1979).

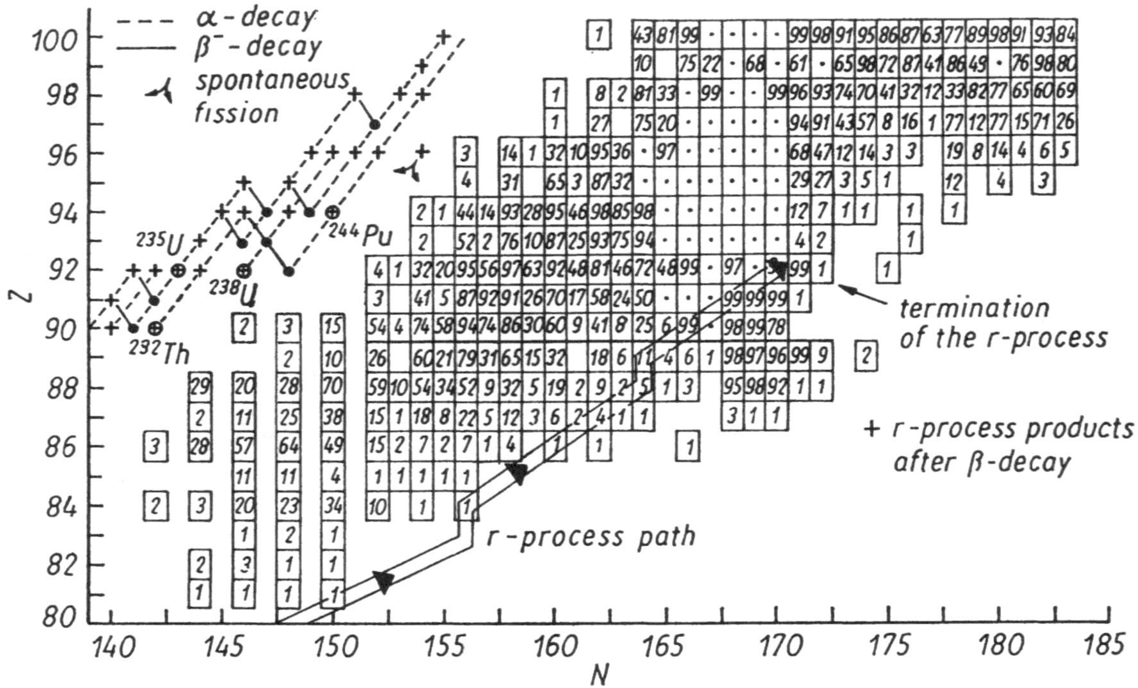

Fig. 17. Illustration of the influence of beta-delayed fission (rates given in %, dots denote 100%) on the r-process path and on the production rates of cosmochronometers. The r-process is prevented by β-delayed fission from producing β-stable elements with $Z \gtrsim 100$. The production ratios of chronometric pairs are dependent on β-delayed fission in the r-path and during decay back to the β-stability line (from Thielemann et al. (1983).)

When denoting by Δ the time interval of nucleosynthesis of r-nuclei in the galaxy until isolation of the presolar cloud, the age of the universe is

$$T_U = (10^9 + \Delta + t_{\text{solar}} + \delta)\ \text{a} \approx (\Delta + 5.7 \cdot 10^9)\ \alpha. \tag{15}$$

Here $t_{\text{solar}} = 4.6 \cdot 10^9$ a is the age of the solar system since its condensation and $\delta \approx 10^8$ a is the time between isolation of the presolar cloud and its condensation. $1 \cdot 10^9$a in eqn. (15) stands for the assumption that the synthesis of r-elements in stars began not later than 10^9 a after the big bang. This is probably an upper limit since the time of decoupling of matter and radiation, marking the onset of formation of galaxies, is estimated to be $\sim$700000 a after the big bang, the prenuclear burning of heavy stars (Kelvin-Helmholtz time) is about 10^7 a and the nuclear burning of a $25 M_\odot$ star until the gravitational collaps is $\sim 8 \cdot 10^6$ a. The assumption of $1 \cdot 10^9$ a thus still leaves time for the formation of supermassive objects which may have produced the first heavy elements which then are further enriched in the r-process of the foregoing section. The nuclei whose abundances probe the quantity Δ and thus the age of the galactic, disk, must have half-lives $T_{1/2} \gg \delta$. The best candidates are ^{238}U, ^{232}Th, and ^{187}Re. With their half-lives of $4.47 \cdot 10^9$ a, $1.405 \cdot 10^{10}$ and $5 \cdot 10^{10}$ a, respectively, they should receive substantial contributions from the entire duration of nucleosynthesis and their abundances are practically independent of δ. (A problem in the use of the ^{187}Re—^{187}Os chronometer is, however, that the neutron capture cross-sections ratio of the osmium isotopes $\sigma_{186}/\sigma_{187}$ must be known for stellar s-process temperatures (Symbalisty and Schramm (1981)). The shorter lived r-only nuclei ^{129}I and ^{244}Pu ($T_{1/2} = 1.57 \cdot 10^7$ and $8.26 \cdot 10^7$ a, respectively) are probes of last local synthesis events which may have triggered the formation of the solar system and are sensitive to the magnitude of δ. ^{235}U (with $T_{1/2} = 7.04 \cdot 10^8$ a) takes an intermediate position.

The quantity, out of which in principle Δ is extracted is

$$R_{ij} = \frac{P_i/P_j}{N_i/N_j}, \tag{16}$$

where P_i/P_j is the ratio of production rates of two chronometers in the r-process and N_i/N_j is their abundance ratio at the time of condensation of the solar material. In principle one needs further a model of the galactic evolution. Different models give a different time dependence of the production function; however, the observational constraints are not very stringent. For discussions of chemical evolution models see e.g. Hainebach and Schramm (1977), Tinsley (1975, 1980), Audouze and Tinsley (1976), Pagel and Edmunds (1981). Because of the large number of synthesizing events (supernova explosions) and mixing by the galactic rotation, often continuous nucleosynthesis is assumed which is either constant over its duration Δ or has decreased monotonically with time, in this way schematically taking into account the condensation of material in white dwarfs, neutron stars and black holes which then will no more participate in synthesizing processes. A simple exponential model (Fig. 18) of galactic nucleosynthesis assumes (Fowler and Hoyle (1960), Fowler (1972, 1978)) a continuous nucleosynthesis rate decreasing with a time constant T_R ($T_R = \infty$ would correspond to a constant production function) over the time Δ and takes into account explicitely a last production spike corresponding to the last passage of a spiral arm, which contributes the fraction S to the total galactic nucleosynthesis, followed then by the free decay period δ. The parameters Δ, S, T_R and δ can then be determined from a system of four non-linear equations (see Thielemann, Klapdor, Metzinger (1983)) into which the r-production ratios of four chronometric pairs enter as well as their abundance ratios at time $\delta + \Delta$ (from meteoritic studies). It would seem reasonable to add not only one but several final spikes to the exponential model (as done by TRIVEDI (1977)). A solution for such a model would need, however, the safe knowledge of more chronometric pairs.

The central problem of the method of cosmochronology by the actinide chronometers was up to now that the reliability of the method depends crucially on the production rates of these nuclides in the r-process which enter into the expression (16). These r-process production ratios were estimated up to now, in most cases, in an extremely crude way following the procedure of Burbidge et al. (1957) and SEEGER and SCHRAMM (1970). A uniform filling of all atomic masses in the actinide region is assumed, and the relative production rates are taken as the ratio of the number of their α-decaying progenitors (SYMBALISTY and SCHRAMM (1981)). This procedure ignores that the β-decay from the r-process path to the stability line, which feeds the α-progenitors, runs through the region of β-delayed fission (Fig. 17). It ignores further the depletion by β-delayed fission

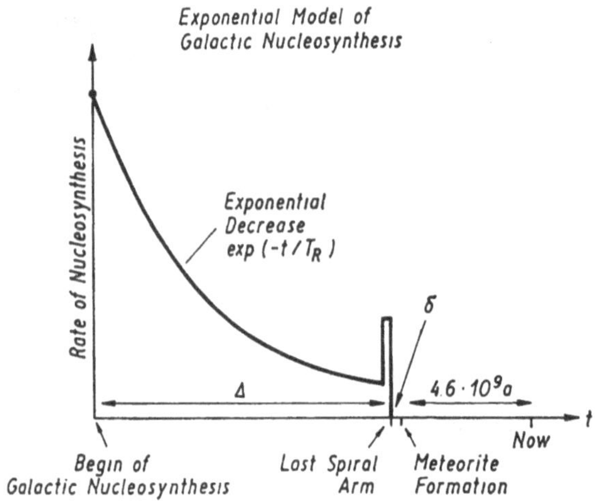

Fig. 18. Illustration of the exponential model for galactic nucleosynthesis used to determine the age of the galaxy from chronometric pairs (see text)

in the r-process path itself. Both effects will change the hitherto adopted "standard" production rates. Already the schematical treatment of beta-delayed fission (with $S_\beta \sim \varrho(E)$ and a one-hump parametrization of the fission barrier) during the decay back to the stability line (WENE (1975)) reduced the P_i/P_j value for ^{232}Th/^{238}U from the "standard value" of 1.9 (SCHRAMM (1974) and SYMBALISTY and SCHRAMM (1981)) to 1.5—1.7.

We find (THIELEMANN et al. (1983)), starting with the element distribution in the r-process path as it results in the scenario of explosive He-burning (see foregoing section) using the microscopically calculated ratios for beta-delayed fission in the r-path and during decay back to the stability line and taking into account β-delayed neutron emission, production ratios of the ^{232}Th/^{238}U, ^{235}U/^{238}U and ^{244}Pu/^{238}U chronometric pairs of 1.39, 1.24 and 0.12, respectively, i.e. they are all remarkably lower than the adopted "standard" values of 1.9, 1.5 and 0.9, respectively, reported by SYMBALISTY and SCHRAMM (1981). The production ratio of ^{247}Cm/^{238}U we find to be $\sim 10^{-2}$.

The calculated ratios are compared with earlier calculations or assumptions in Table 2. It should be noted here, that for the calculation of the production ratios of the chronometric pairs the question of the astrophysical site of the r-process, discussed in the previous section, is of minor importance. Important is only the agreement obtained between the solar r-abundances and the calculated abundance curve in Fig. 15 which reflects two facts:

the location of the r-process path in the $N - Z$ plane and the abundance distribution *in* the r-process are correctly calculated. Thus this abundance distribution, from which the production ratios of the chronometric pairs are deduced, has to be produced by *any* event which can reproduce solar r-abundances.

How do the changes in the production ratios of the cosmochronometers influence the deduced age of the galaxy? In the above mentioned exponential model the use of the four chronometric pairs $^{232}Th/^{238}U$, $^{235}U/^{238}U$, $^{244}Pu/^{238}U$ and $^{129}I/^{127}I$ with the observed abundance ratios at the time of the condensation of the solar system as given in Table 3

Table 2
r-process production ratios of the actinides from different cosmochronological work

$^{232}Th/^{238}U$	$^{235}U/^{238}U$	$^{244}Pu/^{238}U$	Work
1.65	1.65		FOWLER and HOYLE (1960) estimates
1.90	1.89	0.96	SEEGER and SCHRAMM (1970)
1.65	1.42	0.90	FOWLER (1972, 1978)
1.70	0.89	0.53	WENE and JOHANSSON (1976)
1.50	1.10	0.40	KRUMLINDE, et al. (1981)
$1.9^{+0.2}_{-0.4}$	$1.5^{+0.5}_{-0.6}$	$0.9^{+0.1}_{-0.2}$	Adopted standard, SYMBALISTY and SCHRAMM (1981)
1.39	1.24	0.12	THIELEMANN, METZINGER, KLAPDOR (1983)

Table 3
Abundance ratios of chronometric pairs at the time of the condensation of the solar system deduced from meteorite studies

$\frac{\text{nucleus } i}{\text{nucleus } j}$	$\frac{Y_i(\Delta + \delta)}{Y_j(\Delta + \delta)}$		
$\frac{^{232}Th}{^{238}U}$	2.50 ± 0.2	See	SYMBALISTY, and SCHRAMM (1981)
$\frac{^{235}U}{^{238}U}$	0.313 ± 0.026	See	SYMBALISTY and SCHRAMM (1981); BEGEMANN (1980); CHEN, and WASSERBURG (1980)
$\frac{^{244}Pu}{^{238}U}$	0.005 ± 0.002	See	HUDSON, et al. (1982); Marti, et al. (1977)
$\frac{^{129}I}{^{127}I}$	$(0.8 \cdots 2.3) \times 10^{-4}$	See	JORDAN, et al. (1980)

(and with an r-process production ratio for $^{129}I/^{127}I$ of 1.26 from the empirical r- and s-decomposition by KÄPPELER et al. (1982)) results in the following model parameters (THIELEMANN et al. (1983)): $\Delta = (16.1^{+2}_{-4} \cdot 10^9\,a$, $S = 0.057$, $T_R \cong \infty, \delta = 0.162 \cdot 10^9\,a$. Of these values particularly T_R is very sensitive to the value of the observed $^{244}Pu/^{238}U$ abundance: it decreases to the rather improbable value of 4.80 already for an $^{244}Pu/^{238}U$ ratio of 0.008. (According to MILLER and SCALO (1979) the stellar birth rate has not increased by more than a factor of 3 or decreased by more than a factor of 5 over the age of the galaxy.) It is further found that the system of equations gives a solution with the older meteoritit values for the $^{244}Pu/^{238}U$ pair not properlycorrect ed for fractionation effects (0.035 to

0.016, WASSERBURG et al. (1969), PODOSEK (1972), DROZD et al. (1977)) only when the fission barriers are systematically enhanced by 0.5—0.8 MeV. Table 4 gives a summary of the results. Use of the more recent $^{232}Th/^{238}U$ value 2.32 of ANDERS and EBIHARA (1982) would result in a value of Δ close to that obtained with the Symbalisty-Schramm value and $^{244}Pu/^{238}U = 0.006$.

Table 4

Parameters for the exponential model of galactic evolution derived from the r-process production ratios calculated by THIELEMANN, et al. (1983) and the abundance ratios of the chronometric pairs $^{232}Th/^{238}U$, $^{235}U/^{238}U$, $^{244}Pu/^{238}U$, and $^{129}I/^{127}I$ at the time of condensation of the solar system as deduced from meteoritic studies

Model	Assumed abundance ratios at $\Delta + \delta$		Δ*	S	T_R*	$2\times\tau_{SE}^*$	δ*
	$^{244}Pu/^{238}U$	$^{129}I/^{127}I$					
	0.004	—	16.16	0.062	∞	0.169	
	0.005	—	16.10	0.057	∞	0.162	
r-process calculation 1	0.006	1.00	13.93	0.066	21.74	0.153	10.94
	0.007	1.40	12.16	0.078	8.77	0.149	
	0.008	1.80	10.77	0.090	4.80	0.146	
2	0.030	1.07	11.54	0.151	3.04	0.169	11.64
3	0.035	0.90	14.24	0.144	5.42	0.172	12.21

* In units of 10^9 a: Δ (duration of nucleosynthesis), S (part of nucleosnythesis in the last spiral arm), T_R (exponential decrease of nucleosynthesis), δ (free decay period after passage of the last spiral arm), τ_{SE} (age of the elements for nucleosynthesis in a single event deduced from the $^{232}Th/^{238}U$ ratio); 1: fission barriers from HOWARD, and MØLLER (1980); 2,3: first barrier increased by 0.5 or 0.8 MeV, respectively.

Including the uncertainty in the fission barrier heights and in the observed meteoritic ratios there results for the age of the galaxy

$$T_G = (\Delta + \delta + 4.6) \cdot 10^9 \text{ a} = (20.8^{+2}_{-4}) \text{ a}$$

and for the age of the universe

$$T_U = (21.8^{+2}_{-4}) \cdot 10^9 \text{ a}.$$

For comparison the value obtained for the quantity $2\tau_{SE}$ may be given where τ_{SE} is the free decay time needed to transform the $^{232}Th/^{238}U$ production ratio into the ratio observed at time $\Delta + \delta$ assuming a single production event. This value is found to be $2\tau_{SE} = 11.0 \cdot 10^9$ a. It allows, for a restricted set of galactic models (see TINSLEY (1975), SYMBALISTY and SCHRAMM (1981)), a rough rather model independent estimate of the time Δ or its lower limit because sufficiently long-lived chronometers ($T_{1/2} \gg T_G$, which is not well fulfilled, however, for the $^{232}Th/^{238}U$ pair) see the production period essentially as one event occuring in the middle of the production time interval.

The new age of the galaxy of $T_G = 20.8 \cdot 10^9$ a is larger than the values extracted from the actinide chronometers in earlier studies by almost a factor of 2. Earlier values were $T_G = (11.6 \pm 2) \cdot 10^9$ a (FOWLER (1972)) and $T_G = (10.8 \pm 2.3) \cdot 10^9$ a (FOWLER (1978)). It is also larger than Re/Os ages for which recent values are $(16.9 \pm 4) \cdot 10^9$ a (SYMBALISTY and SCHRAMM (1981)), $15 \cdot 10^9$ a (WINTERS et al. (1980)), $15.5 \cdot 10^9$ a (BROWNE and BERMAN (1981)), $13.6 \cdot 10^9$ a (WINTERS and MACKLIN (1982)). The recent study by

YOKOI et al. (1983) confirms the rather large present uncertainty of this chronometer which mainly originates from the temperature- and density-dependent half-life of ^{187}Re. Of comparable uncertainty is at present the ^{176}Lu s-process cosmochronometer (see BEER et al. (1981)) from which BEER and KÄPPELER (1980) derived in a first estimate $\Delta = 9.5 \cdot 10^9$ a. In all these analyses, however, a ratio $T_R/\Delta = 0.43$ was used following the earlier findings of FOWLER (1972, 1978), which is not consistent with the time dependence of the effective nucleosynthesis rate for presently discussed evolution models. Taking $T_R \cong \infty$ would shift the value $\Delta = 9.5 \cdot 10^9$ a by BEER and KÄPPELER (1980) to $\Delta \cong 13 \cdot 10^9$ a and the result for the ^{187}Re/^{187}Os pair would scale similarly to $\Delta \cong 15 \cdot 10^9$ a.

Much better is the agreement of the new age of the galaxy with recent observations. While older studies of the age of globular clusters gave values around $(13 \text{ to } 15 \pm 3) \cdot 10^9$ a (IBEN (1968, 1974)) recent observations by NISSEN (1982) yield e.g. for the globular cluster NGC 6397 an age of $\sim 20 \cdot 10^9$ a and other recent investigations of globular clusters (SANDAGE (1982)) have resulted in similar high ages. The systematic increase of the observational values of the age of globular clusters results from better observations of the primordial He-abundance in the galaxy and of main sequence turn-off luminosities.

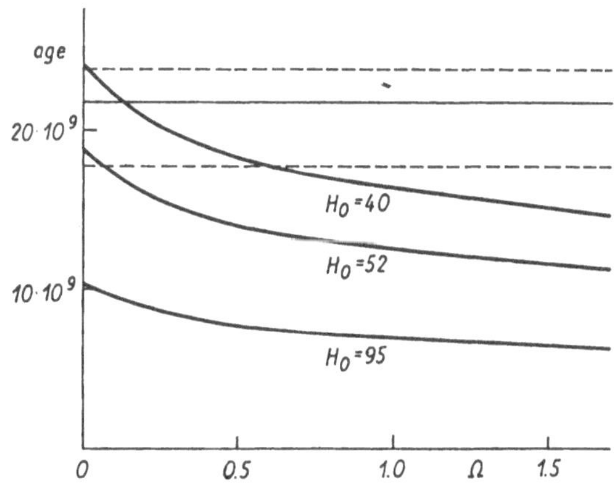

Fig. 19. The age of the universe as function of the density parameter $\Omega = \varrho_0/\varrho_c$ for various values of the Hubble constant H_0. With the global value of $H_0 = (50 \pm 7)$ km s^{-1}Mpc^{-1} of SANDAGE and TAMMANN (1982) the age of the universe deduced from the actinide chronometers (solid line, the dashed lines give the uncertainty) limits the parameter Ω to < 0.6, corresponding to an open universe

It is further of interest, to compare our result with recent determinations of the Hubble constant. SANDAGE and TAMMANN (1982) give as global value $H_0 = (50 \pm 7)$ km s^{-1} Mpc^{-1}. The corresponding upper limit for the age of the universe (the Hubble-time) $T_0 = H_0^{-1} = 19.5^{+3.2}_{-2.4}$ (age for extreme Friedman models with deceleration parameter $q_0 = 0$) is consistent with the actinide age. We can argue, however, also in the following way. Friedman models of the universe with cosmological constant $\Lambda = 0$ are determined by two parameters, the Hubble constant H_0 and the present mass density ϱ_0 of the universe. Thus the age of the universe can be represented, for given H_0, as function of $\Omega = \varrho_0/\varrho_c = 2q_0$, where ϱ_c is the critical mass density, beyond which the universe would be closed. The new actinide age would in such a representation, for the above value of H_0, restrict the parameter Ω to a value < 0.6 (Fig. 19) indicating an open universe. Comparing the corresponding mass density ϱ_0 with the lower limit of the observed baryonic mass density $\varrho_{obs} = (0.7 \pm 0.5) \cdot 10^{-30}$ gcm^{-3} (PRIESTER (1984)) and interpreting the difference as totally due to a finite neutrino mass m_ν would lead, with an assumed neutrino density of 330 cm^{-3}, to an upper limit of $m_\nu \lesssim 4.4$ eV (see also KLAPDOR (1984, 1984 a)).

3.1.3. Galactic neutrinos

Since more than a decade solar (and galactic) neutrinos are detected in the famous ^{37}Cl experiment (DAVIS et al. (1968, 1978)). There are four major neutrino branches that should be measured in a program of neutrino spectroscopy of the solar interior: the pp, ^{7}Be, ^{8}B, and ^{13}N + ^{15}O neutrinos (see BAHCALL (1978)). Spectroscopy of galactic neutrinos can give information on gravitational collapses of stars. The efficiency and its dependence on the incident neutrino energy is, for all neutrino detectors under discussion,

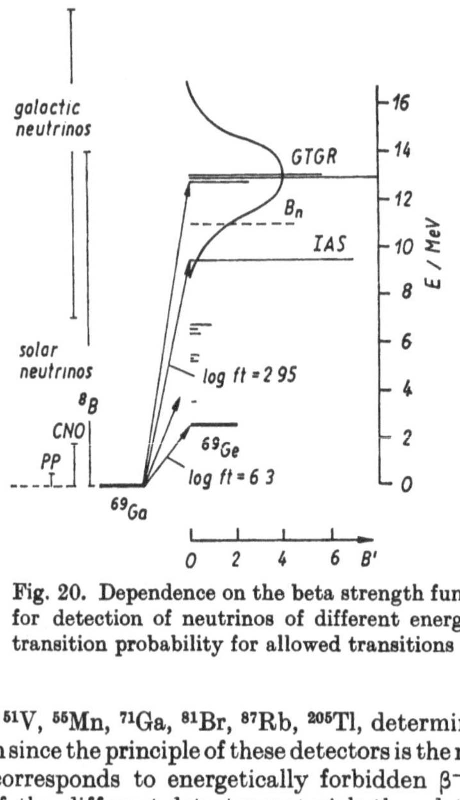

Fig. 20. Dependence on the beta strength function of the efficiency of the ^{69}Ga detector for detection of neutrinos of different energy. Shown is the calculated reduced beta transition probability for allowed transitions (from KLAPDOR and METZINGER (1982 c))

such as ^{51}V, ^{55}Mn, ^{71}Ga, ^{81}Br, ^{87}Rb, ^{205}Tl, determined by the shape of the beta strength function since the principle of these detectors is the neutrino capture reaction $_ZA(\nu_e, e^-)_{Z+1}A$ which corresponds to energetically forbidden β^- decay of $_ZA$ (see Fig. 20). The shape of S_{β^-} of the different detector materials thus determines the degree of applicability of these detectors to the different problems (KLAPDOR (1982, 1983 a)). Up to now there exist, however, for the different detectors only rough estimates (BAHCALL (1978, 1982)) of the contribution of ν capture by excited states to the capture cross-sections.

Let us consider the possibility of detecting stellar collapses by the emitted neutrino burst. Assuming that the collapse of a heavy star leads to the emission of $\sim 2 \cdot 10^{53}$ erg in form of ν_e and $\sim 2 \cdot 10^{52}$ erg in form $\bar{\nu}$ (BROWN (1977)), WILSON (1980), BETHE et al. (1980), SCHRAMM (1980), SALPETER an SHAPIRO (1981)), leads during less than a second to a time-integrated flux on the earth of the order of $\sim 10^{12} \nu_e$/cm^2, when the collapse occurred in a distance of ~ 10 kpc, i.e. near the center of the galaxy. The expected neutrino energies (which can escape the supernova) are lying between 10 and 20 MeV (Ro-

BERTS et al. (1977), BROWN (1977), ARNETT (1977), WILSON (1971, 1980), BETHE et al. (1980)). Some estimate of the absorption cross-sections of these neutrinos by ^{71}Ga, ^{7}Li, ^{37}Cl has been made by BAHCALL (1978). A discussion is also given by HAMPEL (1982). Bahcall takes into account, in the case of ^{71}Ga, only three excited states in ^{71}Ge, to be compared with the distribution of the main GT strength according to microscopic calculations (KLAPDOR and METZINGER (1982c)) shown in Fig. 20. The summed GT strength lying below the isobaric analogue state in ^{71}Ge we find to correspond to a log $ft \approx 3.45$ (log $ft \approx 3.53$ for ^{69}Ge), to be compared to the log $ft = 6$ and 5, respectively, assumed by BAHCALL (1978) for his considered three states. For the strength concentrated in the GTGR (not considered by Bahcall) we find (assuming a normalization by a factor of 2) a log $ft \approx 2.7$ in ^{71}Ge and about the same value for ^{69}Ge. Fig. 21 shows the difference in

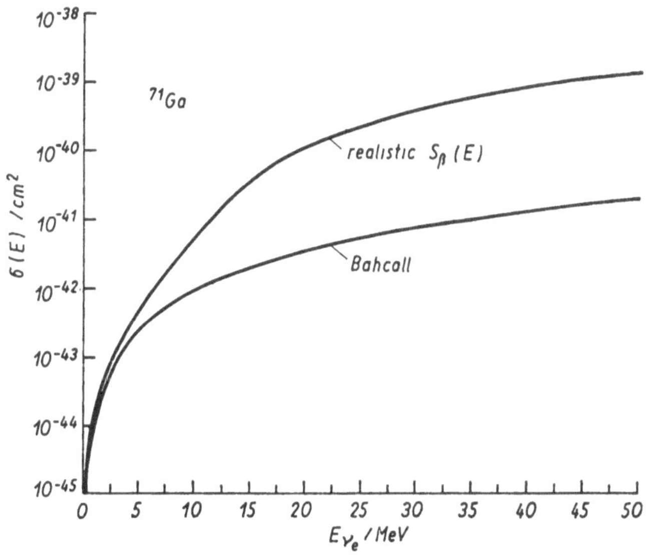

Fig. 21. Capture cross-section for neutrinos by ^{71}Ga as function of the incident neutrino energy under different assumptions on the beta strength function (from KLAPDOR and METZINGER (1982c))

the predicted ν capture cross-sections for ^{71}Ga as function of neutrino energy for the beta strength function used by BAHCALL (1978) and the more realistic microscópic one.

^{71}Ga seems compared to ^{69}Ga, less useful as a detector for galactic neutrinos, because most of the strength is concentrated beyond the neutron binding energy in ^{71}Ge, leading to a subsequent decay to the stable ^{70}Ge. HAMPEL (1980) has pointed out the advantage of the ^{69}Ga detector of being almost free of a solar neutrino background. It was assumed, however, that because of the short life-time of ^{69}Ge($T_{1/2} = 39$ h) and the decay of the ^{69}Ge signal after a few days, a trigger signal (e.g. from the Homestake Long Range Neutrino Detector) would be needed. This problem may not exist when the existence of the GTGR is taken into account (KLAPDOR (1982)). From the latter subsequent neutron emission leads to ^{68}Ge with $T_{1/2} = 288$ d. The percentage of incident neutrinos from a stellar collapse captured in the GTGR depends on their energy as shown in Fig. 22. For $E_\nu = 20$ MeV about 40% would be captured in the GTGR. This should allow the performance of a galactic neutrino experiment independent of an external trigger simultaneously with the planned solar neutrino Ga experiment, provided a method could be developed to detect the low count rate from ^{68}Ge (or the subsequent decay of ^{68}Ga).

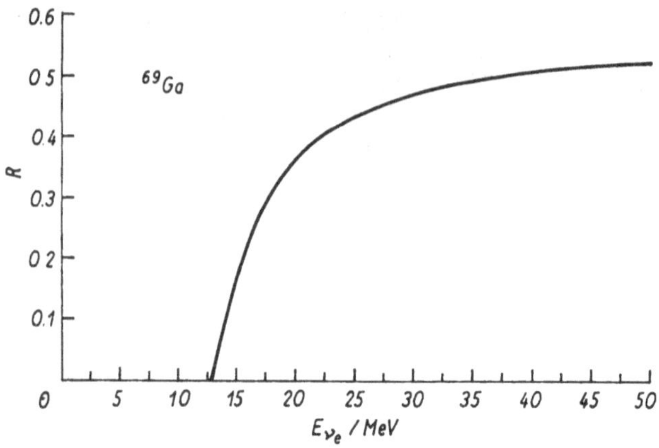

Fig. 22. Ratio R of the cross-section for neutrino capture in the Gamow-Teller giant resonance and the total neutrino capture cross-section in ^{69}Ga as function of incident energy (from KLAPDOR and METZINGER (1982c))

3.2. Nuclear Physics

We shall discuss the following of the subjects in nuclear physics, which are sensitive to the shape of the beta strength function.

(1) The determination of fission barrier heights from measured beta-delayed fission branching ratios. This method is the only one allowing the investigation of fission barriers of "cold" nuclei far from the beta-stability line.

(2) The production of transuranium elements using the neutron flux of thermonuclear explosions. Beta-delayed fission influences the production cross-sections.

(3) Double beta decay.

(4) Nucleon density distributions and charge exchange experiments.

(5) The electron and anti-neutrino spectra in nuclear reactors. Their precise knowledge is decisive for the interpretation of reactor neutrino oscillation experiments.

3.2.1. Fission barriers from β-delayed fission

Fission barriers have been investigated, close to the stability line, mainly by direct reactions (BACK et al. (1974a, b); BRITT (1980)) and from fission isomers (BRITT et al. (1973, 1980); METAG et al. (1980)). The possibilities of these methods are presently almost exhausted (BRITT (1980)).

Information on the fission barriers far from stability can in principle be obtained by measuring $\langle \Gamma_n/\Gamma_f \rangle$-values in (HI, xn) reactions. The problem, however, to extract from measured $\langle \Gamma_n/\Gamma_f \rangle$-values (averaged over a wide range of excitation energies) a reliable fission barrier of the cold nucleus is not yet completely solved (see the review paper by LAZAREW et al. (1980)).

For nuclides far from the stability line, the only method to study the barriers of cold nuclei, is β-delayed fission. This has been pointed out already by KLAPDOR et al. (1978a, 1979) and also that a reliable determination of fission barrier heights can be made from measured $P_{\beta df}$ values only when properly taking into account the detailed structure of the beta strength function. Fig. 23 shows an example: The dependence of the fission barrier height extracted from the measured (HABS et al. (1978)) P_{df} value of $1.3 \cdot 10^{-2}$

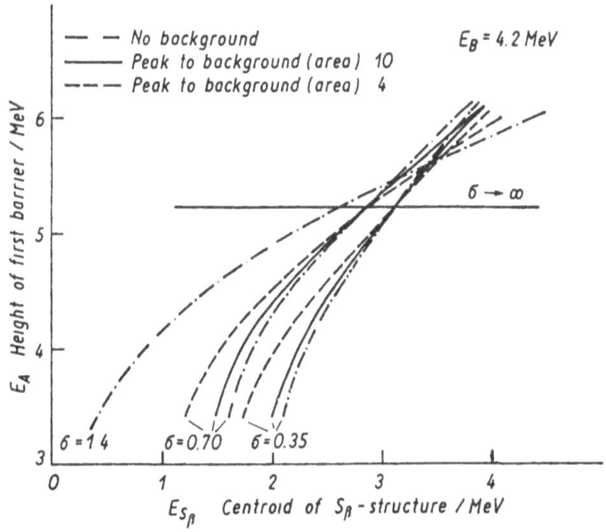

Fig. 23. Dependence of fission barrier heights extracted from measured β-delayed fission branching ratios $P_{\beta df}$ on the shape of the beta strength function. As example the height of the first barrier (E_A) in ^{232}Pu is calculated here from the measured (HABS et al. (1978)) value $P_{\beta df} = 1.3 \cdot 10^{-2}$ for Am $\xrightarrow{\text{EC}}$ ^{232}Pu → fission, as function of the position of a single structure (see Fig. 24) in $S_\beta(E)$. The structure in $S_\beta(E)$ is assumed to have a Gaussian shape. Its width σ is varied as well as the ratio between the strength in the structure and of an assumed background within the Q_β window (which might arise e.g. from forbidden transitions): —·—·— no background; ——— peak to background (area) 10; — — — peak to background (area) 4 (from KLAPDOR and WENE (1980))

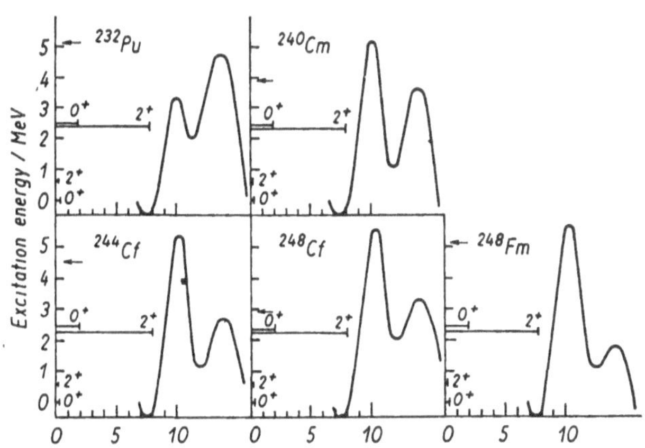

Fig. 24. Calculated (relative) beta strength functions for electron capture to the indicated nuclei (from KLAPDOR et al. (1979)). The fission barriers are from MÖLLER and NIX (1974), the Q_{EC} values estimated from the Garvey-Kelson (1976) mass formula (see JÄNECKE (1976))

for beta-delayed fission of ^{232}Am on the structure of the beta strength function. This example is particularly transparent because of the simple structure of $S_{\beta^+}(E)$, shown for ^{232}Pu and some other actinides in Fig. 24 (from KLAPDOR et al. (1979)). Keeping the second barrier fixed at $E_B = 4.2$ MeV, which is within the interval 4.2—4.5 MeV assumed by HABS et al. (1978) on the basis of isomer data, the height of the first barrier is shown in Fig. 23 as function of the position of the peak in $S_{\beta^+}(E)$ (for the experimentally fixed $P_{\beta df}$ value). The fission barrier penetration has been treated here in a way ignoring transmission resonances. With the structure of $S_{\beta^+}(E)$ given in Fig. 24 and assuming a peak width (FWHM) of about 1 MeV ($\sigma = 0.35$) we extract from the measured $P_{\beta df}$ value a height of the first barrier in ^{232}Pu of $E_A = 4.0$—4.4 MeV, i.e. within the errors of ± 1 MeV quoted (MØLLER and NIX (1973, 1974); HOWARD and MØLLER (1980)) for the Strutinsky procedure. The assumption of a constant beta strength function would have required a higher value for the first barrier and would have led to the erroneous conclusion of a kind of a second "Thorium anomaly" in the fission barriers (HABS et al. (1978)).

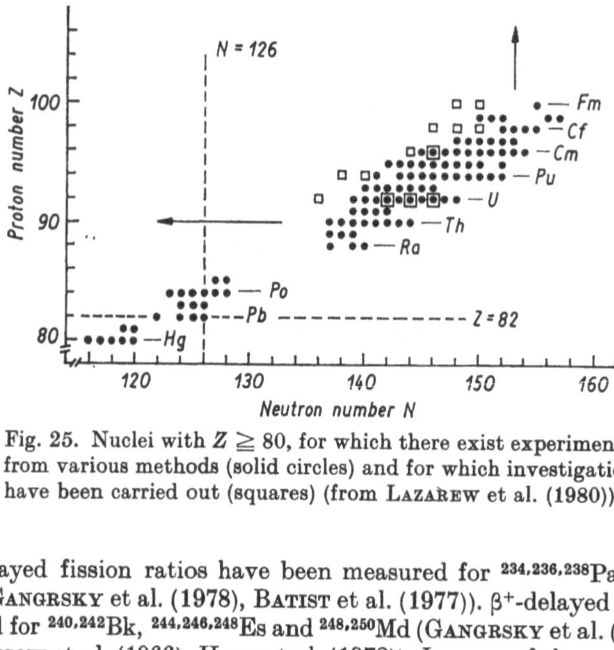

Fig. 25. Nuclei with $Z \geqq 80$, for which there exist experimental data on fission barriers from various methods (solid circles) and for which investigations of beta-delayed fission have been carried out (squares) (from LAZAREW et al. (1980))

β^--delayed fission ratios have been measured for ^{234,236,238}Pa and ^{232}Ac (GANGRSKY (1977), GANGRSKY et al. (1978), BATIST et al. (1977)). β^+-delayed fission ratios have been observed for ^{240,242}Bk, ^{244,246,248}Es and ^{248,250}Md (GANGRSKY et al. (1980)) and for ^{232,234}Am (KUSNETSOW et al. (1966), HABS et al. (1978)). In none of these papers, however, has an analysis of the measured $P_{\beta df}$ values been made by using a proper description of the beta strength function.

Fig. 25 (from LAZAREW et al. (1980)) shows the nuclei with $Z \geqq 80$ for which there exist at present experimental data on fission barriers from β-delayed fission and other methods and may serve, together with Fig. 26, to illustrate the possibilities which may lie in β-delayed fission as a method for the study of the structure of potential energy surfaces associated with fission.

It is well known (and illustrated by Fig. 25) that experimental values of fission barriers exist only over a small range of variation of various of the parameters entering into the usual macroscopic-microscopic description of fission barriers. This has led to the result that different theories can describe "reasonably" the known barriers with substantially different parameters. To give only one example, we mention the isospin dependence of the macroscopic (LD) fission barrier height. For those Thorium nuclei for which the fission barriers are known ($N = 137-144$), drastically different assumptions on the dependence of the macroscopic barrier on T_3 were made in different calculations with a difference in macroscopic barrier heights up to a factor of 2 (LAZAREW et al. (1980)). It is obvious that such uncertainties which can be counteracted by corresponding choice of other parameters in the comparatively narrow (in Z and N) region, for which barriers are experimentally known, must lead to large uncertainties in predictions of fission barriers far from the known region, e.g. for superheavy nuclei. It is, on the other hand, of vital interest for any experimental attempt to synthesize superheavy elements, to know the contours of the superheavy island, in particular the decrease of the fission barrier heights

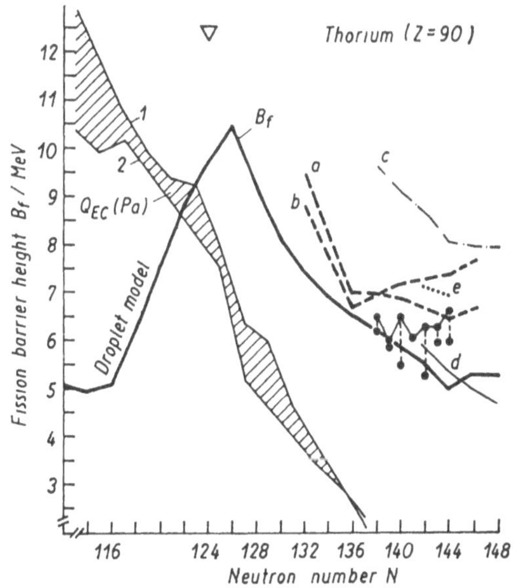

Fig. 26. Experimental (solid circles, from BRITT (1980)) and theoretical values of fission barrier heights for Th isotopes. The solid line denotes the prediction of the droplet model of MYERS (1977), curves a, b, c, d and e and the point ∇ are results of macroscopic-microscopic calculations of B_f^{max} by MØLLER (1975) (curves (a) and (b)), MØLLER (1980) (curve (c)), MØLLER and NIX (1974) (curve (d)), and PAULI and LEDERGERBER (1971) (curve (e)), respectively, the point ∇ is the prediction by ABERG et al. (1980) for ^{214}Th. The shaded area denotes predicted Q_{EC} values for β^+ decay of Pa isotopes according to VIOLA et al. (1974) (curve 1) and MYERS (1977) (curve 2) (from LAZAREW et al. (1980))

when moving away from the doubly magic $^{298}114$. Different calculations here still give quite different answers (see, e.g. FISET and NIX (1972) and RANDRUP et al. (1974)). A unique possibility of getting general information about this point without extending to the region of unknown elements is the study of fission barriers of nuclei lying in N and Z around the closed nucleon shells wich $N = 126$ and $Z = 82$. This is discussed in detail by LAZAREW et al. (1980), who show that an investigation of nuclei ~20 mass units away from the beta stability line in the range $Z \gtrsim 80$, $N \lesssim 126$ by beta-delayed fission should be feasible by heavy-ion reactions. Fig. 26 demonstrates the idea for the neutron-deficient Thorium isotopes. The Q-value for the β-decay of Pa continuously grows with decreasing N, and β-delayed fission will be "switched on" when crossing the $N = 126$ shell, when $Q_{EC} \gtrsim B_f$. The onset of β-delayed fission will depend on the heights of the fission barriers — and on the shape of the beta strength function! A systematic search for those neutron-rich actinides close to the β stability line which are the best candidates for experimental investigation of fission barriers, i.e. for which the largest rates of β^--delayed fission are expected, has been performed by METZINGER and KLAPDOR (1983). The results are given in Table 5.

Table 5

Calculated β^- delayed fission rates $P_{\beta df}$ for those actinides close to the β stability line for which the largest rates of $P_{\beta df}$ are expected (see text). The upper limit results when placing the whole β^- strength 100 keV below the β^--decaying g.s.

nuclide	$P_{\beta df}$ (upper limit)	$P_{\beta df}$
^{244}Np	$1.1 \cdot 10^{-4}$	$1.4 \cdot 10^{-8}$
^{246}Am	$1.1 \cdot 10^{-6}$	$1.1 \cdot 10^{-8}$
^{248}Am	$1.7 \cdot 10^{-4}$	$7.8 \cdot 10^{-7}$
^{252}Bk	$1.5 \cdot 10^{-5}$	$1.2 \cdot 10^{-7}$

3.2.2. Production of transuranium elements by thermonuclear devices

In thermonuclear explosions neutron fluxes of $\gtrsim 10^{31}$cm^{-1}s^{-1} are reached corresponding to target exposures of $\sim 10^{25}$ cm^{-2} within less than 1 μs (BELL (1967)). In the first terrestrial thermonuclear explosion (the "Mike" event of 1952) heavy nuclei were produced by the exposure of ^{238}U to this high neutron flux, up to $A = 255$, the elements Es and Fm were discovered as well as ^{254}Cf (BELL (1967)). A summary of all later explosions up to 1969 is given by INGLEY (1969). A recent review is given by HOFF (1978). There are several features in the production rates of heavy nuclei in these events, for which beta-delayed fission may provide an explanation.

(1) The most productive tests, concerning the production of heavy elements were the CYCLAMEN and HUTCH events (HOFFMAN (1966), ECCLES (1970), HOFF and HULET (1970)) detonated underground in the Nevada test site 1966 and 1969. The HUTCH event has provided by subsequent chemical processing of debris-rich rock larger amounts of ^{257}Fm and ^{250}Cm than had been available before (HOFF (1978)) — which then was used in a number of significant experiments to investigate neutron-induced fission of ^{257}Fm, for the production of ^{258}Fm (HULET et al. (1971), JOHN et al. (1971), WILD et al. (1973)), further the production of ^{251}Cm and the establishment of a level scheme of ^{251}Bk (HOFF (1978)). However, no evidence was found for the production of nuclides beyond ^{257}Fm. The positive result concerning the production of ^{250}Cm has, on the other hand, given rise to a proposal (HULET et al. (1973)) to produce in a HUTCH-like device ^{250}Cm as target material for the search for superheavy elements in heavy-ion reactions.

(2) An "odd-even reversal" in the mass-yield curve was found to occur at $A = 250$. As a consequence of the pairing effect the even-even U nuclides should have smaller neutron capture cross-sections than nearby odd A uranium isotopes. Consequently, yields of the even mass products would be expected to be relatively higher than those of their odd-mass neighbours. That this effect is reversed against the expectation beyond $A = 250$ is clearly seen in the PAR (DORN and HOFF (1965)), BARBEL (Los Alamos Radiochemistry Group 1965) and CYCLAMEN event (HOFF (1978)).

The fact that no nuclides with $A > 257$ have been observed, was discussed as a consequence of an onset of neutron-induced fission (HOWARD and NIX (1974)) or of a decrease in the half-life for spontaneous fission (HULET et al. (1974), MELDNER (1974, 1976)). However, as a result of the same shell effect responsible for the onset of spontaneous fission beyond $Z \approx 94$, $N \approx 166$, β-delayed fission occurs, as shown in Fig. 11. The region of β-delayed fission in the Z, N-plane is naturally more extended than that of spontaneous fission because of the systematically lower "effective" fission barriers in β-delayed fission. It extends below the "bay of shells" (for the latter see BENGTSSON et al. (1974)) and as a consequence we find uranium isotopes with $A > 257$ already decaying to 100% by β-delayed fission (Fig. 11). So it will be the latter effect which is responsible for the up-to-date non-observation of nuclides with $A > 257$.

MELDNER et al. (1976) proposed to circumvent the dilemma, that heavier elements eventually produced in thermonuclear explosions undergo spontaneous fission before any processing can be completed (the "$Z \approx 94$, $N = 166$ region" is reached by β-decay in a time of the order of several sec after the explosion, i.e. much faster than still assumed by MELDNER (1974)), by detecting the isotopes at entrance into the spontaneous fission region by means of the neutrons emitted in the fission process with a neutron-multiplicity counter. They found that detection of masses up to $A = 265$ should be possible provided the total life-time for all β-decays before spontaneous fission in the $Z \approx 94$, $N \approx 166$ region is $\Sigma\tau_\beta \gtrsim 30$ s. The occurence of β-delayed fission will lead, however, to fission (and the fission neutrons) already outside of the spontaneous fission area, i.e. after less β-decay steps. This imposes much more serious time restrictions on the proposed method, which are further increased by the shorter microscopic β-decay half-lives.

The proposal to synthesize superheavy elements by successive neutron bursts from combined thermonuclear macro- and micro-devices (MELDNER (1974)) bases on the assumption that the β-decay half-lives driving the isotopes produced in the first explosion towards the spontaneous fission area are in the range of minutes. The microscopic beta decay half-lives are, however, shorter by a factor of ~100 (KLAPDOR et al. (1981c, 1982)). The path has further been chosen only with regard to neutron-induced and spontaneous fission but not to β-delayed fission. The advantage of using thermonuclear devices with less fission contribution, i.e. neutron bombs, instead of conservative thermonuclear explosions, for the production of heavy elements was noted by KLAPDOR and WENE (1980).

In order to explain the "odd-even reversal", BELL (1967) assumed that the heaviest products are produced in a capture chain with odd Z, e.g. Np or Pa (produced from the U by (d, n) and (n, p) reactions) while the lighter masses are produced in the uranium capture chain. This ansatz, however, cannot give a consistent explanation of this effect and of the results from the TWEED event (1965) in which a ^{242}Pu target was used (see the discussion by WENE and JOHANSSON (1974)).

A natural explanation of the odd-even effect is given, however, by beta-delayed fission during the decay back to the stability line (see Fig. 11, and KLAPDOR and ODA (1980)). Our microscopic calculations in this point support the qualitative conclusion of WENE and JOHANSSON (1974). It might be noted, that in extractions of neutron capture cross-sections from the yield curves in thermonuclear nucleosynthesis experiments, the existence of beta-delayed fission has to be taken into account beyond $A = 250$.

3.2.3. Double beta decay

In the context of grand unified theories (GUT's) besides baryon number (B) violation (proton decay) also lepton number (L) violation should occur. One of the most sensitive tests of lepton number violation is double beta (ββ) decay (PRIMAKOFF and ROSEN (1959, 1981), PONTECORVO (1968)).

Double beta decay, besides strangeness oscillations of the K^0 meson the only studied effect of second order of the weak interaction, may occur in two modes (Fig. 27).

$$\begin{aligned} (Z, A) &\to (Z + 2, A) + e_1^- + e_2^- + \bar{\nu}_1 + \bar{\nu}_2 \quad \text{or} \\ &\to (Z + 2, A) + e_1^- + e_2^- . \end{aligned} \tag{17}$$

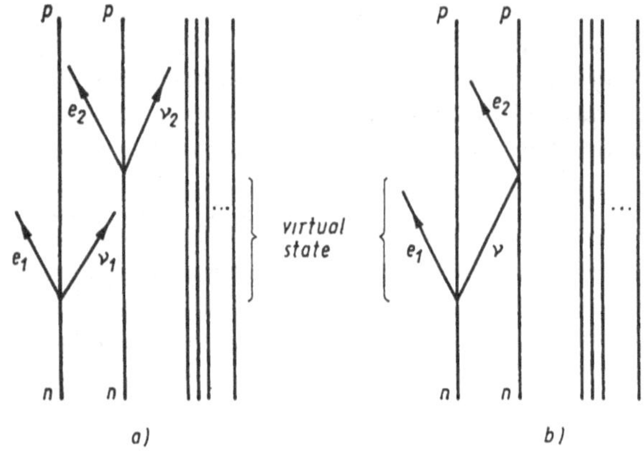

Fig. 27. Modes of double beta decay (a) 2ν decay (b) 0ν decay

The first of these processes conserves lepton number and can be regarded as a sequence of two "normal" β-decays passing through virtual intermediate states of the nucleus ($Z + 1, A$). The second process would be of more fundamental interest, since it would directly indicate a (B–L) violation and could either be interpreted (Fig. 28) in terms of a finite neutrino majorana mass or in terms of a righthanded admixture η in the weak lepton current (Halprin et al. (1976)).

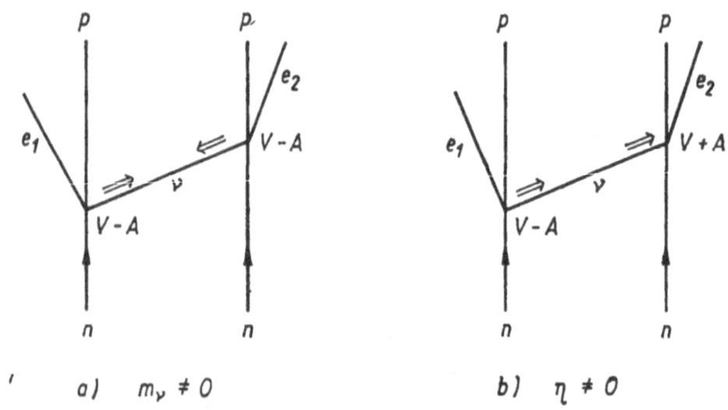

Fig. 28. Two possible implications of an existing 0ν ββ-decay mode. (a) helicity mixing corresponding to a finite neutrino majorana mass (b) right-handed admixture η in the weak lepton current

Although the 0ν mode should demonstrate its existence by a signal with a very distinctive signature, it is very difficult to establish this mode by a direct counting measurement, because of the extremely small decay rates. Experiments up to now yielded only upper limits for such a decay mode (Cleveland et al. (1975), Bardin et al. (1970), Avignone et al. (1983)).

Because of the higher sensitivity achievable, the main interest has been focussed on geochemical experiments (Kirsten and Müller (1969), Srinivasan et al. (1973), Hennecke et al. (1975), Kirsten et al. (1983)) in which the total ββ-decay rates resulting from the contributions of the 2ν and the probably existing 0ν mode are determined. By neglecting contributions from the 2ν mode already upper bounds for m_ν can be obtained. It is believed that more stringent conclusions can be drawn by comparing the ratio of the measured half lives of ^{128}Te and ^{130}Te with the theoretical expectation for the ratio of the 2ν mode (PONTECORVO (1968), KIRSTEN et al. (1983), HAXTON et al. (1981, 1982, 1983), MINKOWSKI (1982), DOI et al. (1983). In fact, *only* this procedure has in principle the power to *establish* the existence of a 0ν mode and thus a finite ν mass or righthanded admixture from geochemical experiments.

The point that shall be made here, is the following:
Calculations of 2νββ decay believed to be of sufficient precision to allow a meaningful comparison between theory and experiment are given by HAXTON et al. (1981, 1982, 1983). However, we feel that the analyses of geochemical results using these calculations may be questionable due to the fact that the latter predict a 2ν ββ-rate for ^{130}Te a factor ~160 higher than one deduces from the geochemical result for the *total* decay rate, while the calculated rate *ratio* for both isotopes has to be known to an accuracy of the order of 50%. But not only for getting the 0ν ββ-rate out of the geochemical results the problems in the shell model calculations existing so far are a severe shortcoming, but also in relating the 0ν ββ-rate to the lepton number violation parameters $\langle m_\nu \rangle_{\text{maj.}}$ and η. Here enters (DOI et al. (1983)) the *absolute* value of the nuclear matrix element $M^{0\nu}$, which certainly cannot be expected to be better known from the calculations than the matrix element for the 2ν ββ-decay and which is usually simply 'scaled' according to $M^{0\nu} = M^{2\nu}_{\text{exp}} \cdot (M^{0\nu}_{\text{theor.}}/M^{2\nu}_{\text{theor.}})$.

In an attempt to understand this discrepancy between theory and experiment we have started to investigate the 2ν ββ decay of ^{128,130}Te, ^{82}Se and ^{76}Ge avoiding the closure approximation used by these authors and in most other calculations up to now.

In our more exact treatment not simply the matrix element of $\sigma\sigma\tau^-\tau^-$ between initial and final 0^+ states is calculated, but matrix elements $\langle M\| \sigma\tau^- \|I\rangle$ and $\langle F\| \sigma\tau^- \|M\rangle$ are calculated for the whole spectrum of intermediate 1^+ states M. Steps into this direction were done by HAXTON et al. (1981) and by HUFFMAN (1970). While in the first work an explicit summation restricted to 1^+ states within only the first 5 MeV excitation energy in ^{82}Br has been performed (which means that the distribution of GT strength in the intermediate nucleus is not described properly since the existence of a GTGR and of a quenching of GT strength by ΔN^{-1} excitation is ignored), in the second the 1^+ spectrum in ^{128,130}I was calculated in an RPA approach, however, treating the lepton phase space approximately. In our calculations, extending a recent study (GROTZ, KLAPDOR, METZINGER (1983)), we calculate the spectrum of intermediate 1^+ states, up to ~300 MeV (KLAPDOR and GROTZ (1983, 1984)).

Starting from time dependent perturbation theory the 2ν ββ $0^+ \to 0^+$ decay rate neglecting contributions from Fermi decay is given by:

$$\omega_{2\nu} = \frac{F_A{}^4 G^4}{32\pi^7} \int\limits_{m_e}^{t_0+m_e} F(Z, \varepsilon_1)\, |K_1|\, \varepsilon_1 \,d\varepsilon_1 \int\limits_{m_e}^{t_0+2m_e-\varepsilon_1} F(Z, \varepsilon_2)\, |K_2|\, \varepsilon_2 \,d\varepsilon_2 \int\limits_0^{t_0-\varepsilon_1-\varepsilon_2+2m_e} \nu_1^2 \nu_2^2 \,d\nu_1 \sum_{MM'} A_{MM'} \tag{18}$$

with $$A_{MM'} = \langle F|\, |\sigma\tau^-|\, |M\rangle \langle M|\, |\sigma\tau^-|\, |I\rangle \langle F|\, |\sigma\tau^-|\, |M'\rangle \langle M'|\, |\sigma\tau^-|\, |I\rangle \times \frac{1}{3} \left(K_M K_{M'} + L_M L_{M'} + \frac{1}{2} K_M L_{M'} + \frac{1}{2} L_M K_{M'} \right)$$

with the same notation as in GROTZ et al. (1983).

The absolute values of $\langle M|\, |\sigma\tau^-|\, |I\rangle$ can in principle be compared to (p, n) charge exchange experiments, from which the β^- Gamow Teller (GT) strength function $B^-_{\text{GT}}(E)$ can be extracted. The matrix elements $\langle F|\, |\sigma\tau^-|\, |M\rangle$ are related to the β^+ strength B^+_{GT} of the final 0^+ ground state by $B^+_{\text{GT}} = |\langle F|\, |\sigma\tau^-|\, |M\rangle|^2$, which could in principle be extracted from an (n, p) experiment with the daughter nucleus as target.

β^- as well as β^+ matrix elements were calculated using PBCS wave functions. In this approximation it is not necessary to truncate the shell model basis seriously. Particle number projection has been done and is necessary because of the large overlap of the initial and final 0^+ ground states. The residual GT interaction $V_{GT} \cdot \sigma\sigma\tau^-\tau^-$ is diagonalized in a 2 q.p. basis for the 1^+ states and a 4 q.p. basis for the 0^+ states. First order ground state correlations (g.s.c.) of the GT interaction are included in this treatment.

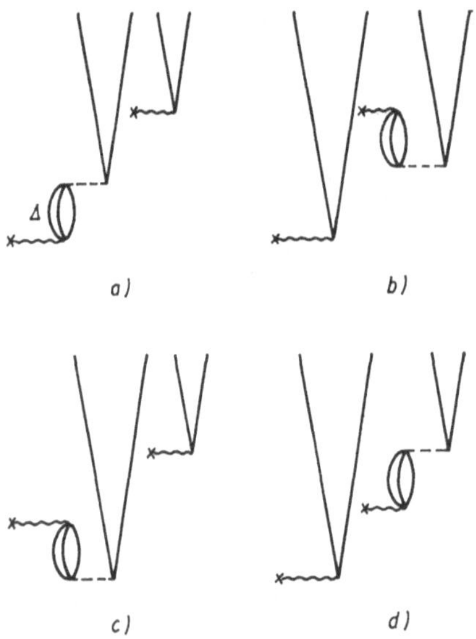

Fig. 29. Interaction graphs including a Δ-excitation which give a sizeable contribution to the $\beta\beta$-amplitudes. In the final state the 4 quasi particles have to couple to the 0^+ ground state which is equivalent to the quasi particle vacuum. Hence graphs c) and d) correspond to g.s. correlations in the initial and final nucleus, respectively

ΔN^{-1} excitations are included in the form of two collective states $\langle \Delta - h, T = 1, T_Z = +1$ and $\langle \Delta - h, T = 1, T_Z = -1$, neglecting the energy splitting of the different Δ-orbits (see GROTZ, KLAPDOR, METZINGER (1983b) and KLAPDOR (1983b)). G.s.c. as well as final state interaction graphs have been included (see Fig. 29). The following results are obtained (KLAPDOR and GROTZ (1983)): The ^{130}Te β^- as well as the ^{130}Xe β^+ amplitudes in this model are shown in Fig. 30. The combined 2ν $\beta\beta$-amplitudes entering in eq. (18) are shown in Fig. 31. The transitions directly via the collective $\Delta - h$ states cancel each other, so that there is no contribution to the $\beta\beta$-decay from the 300 MeV region. The quenching of the 2ν $\beta\beta$ rates by the Δ excitations is surprisingly small (26—29%) compared with the 30% decrease obtained for the *total* β^- strength of the parent nuclei (and smaller than the estimate by ZAMICK and AUERBACH (1982)). The reasons for this are the following: It was shown (GROTZ, KLAPDOR, METZINGER (1983)) that the quenching of β^- strength below the GTGR is much smaller than for the GTGR itself due to a destructive interference of first and second order effects. Secondly, it was found that the quenching of β^+ strength, responsible for the second decay step, is generally weaker. This is because the β^+ strength is already strongly quenched by nucleonic g.s.c. and switching on the $\Delta - h$ excitations reduces this effect at the same time.

As can be seen from Table 6 the discrepancy between calculated and observed decay rates for ^{128}Te and ^{130}Te cannot be removed in this model. The same is true for ^{82}Se. It is

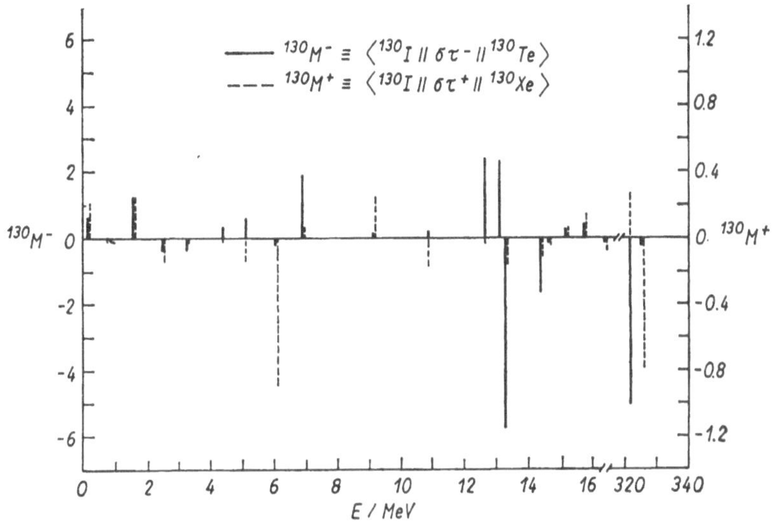

Fig. 30. Calculated Gamow-Teller reduced matrix elements for β^- decay of ^{130}Te and β^+ decay of ^{130}Se. In the calculation single particle level set II (s.p.l. II) of GROTZ, KLAPDOR, METZINGER (1983) was used

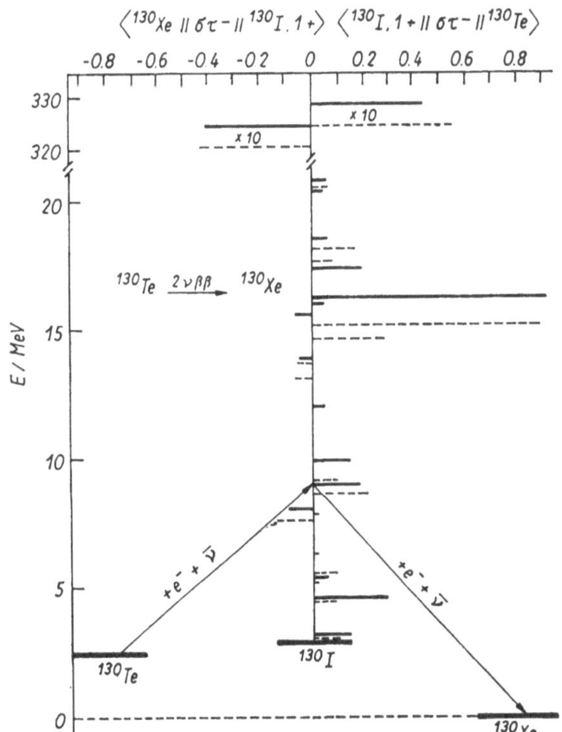

Fig. 31. 2ν ββ-amplitudes for the transition $^{130}\mathrm{Te} \xrightarrow{\beta\beta} {}^{130}\mathrm{Xe}$ deduced from the reduced matrix elements shown in fig. 30 (solid lines) and from a calculation using s.p.l. I of GROTZ, KLAPDOR, METZINGER (1983) (dashed lines)

of interest to note that a lower limit for the 2ν ββ decay rates of ^{128,130}Te and ^{82}Se can be estimated *independently* of nuclear structure calculations. One can calculate the contribution from the lowest intermediate 1^+ state assuming equality of the $\beta^{\pm}$ decay matrix elements for this state with the ones *observed* in the g.s. decays of ^{128}I and ^{80}Br, respectively. Including only this first transition the result of eq. (18) comes very close to the observed ββ-half lives (see Table 6).

Table 6

Calculated (KLAPDOR and GROTZ (1984)) 2νββ half lives, neglecting (first column)and including $\Delta - h$ excitations. The third column shows the effect of quadrupole correlations. The forth column uses only the experimental ground state matrix elements as described in the text. The experimental values in the last column correspond to the sum of 2ν and 0ν decay from the geochemical results, to 2ν decay for ^{76}Ge and the second ^{82}Se value.

	calculated half lives from 2ν ββ-decay in years				experimental half lives in years
	no $\Delta - h$ mixing	with $\Delta - h$ mixing	with $\Delta - h$ mixing and quadrupole-quadrupole correlations	from 1^+ ground state decay	
^{128}Te	$6.2 \cdot 10^{22}$	$8.7 \cdot 10^{22}$	$5.7 \cdot 10^{23}$	$1.60 \cdot 10^{25}$	$> 8 \cdot 10^{24}$ (2σ) Kirsten et al. (1983) $(1.5 \pm 0.2) \cdot 10^{24}$ HENNECKE et al. (1975)
^{130}Te	$1.5 \cdot 10^{19}$	$2.0 \cdot 10^{19}$	$1.2 \cdot 10^{20}$	$2.55 \cdot 10^{21}$	$(2.55 \pm 0.20) \cdot 10^{21}$ KIRSTEN (1982)
^{82}Se	$5.1 \cdot 10^{18}$	$7.2 \cdot 10^{18}$	$1.5 \cdot 10^{18}$	$1.81 \cdot 10^{20}$	$(1.45 \pm 0.15) \cdot 10^{20}$ KIRSTEN (1982) $(1.0 \pm 0.4) \cdot 10^{19}$ MOE and LÖWENTHAL (1980)
^{76}Ge	$1.3 \cdot 10^{20}$	$1.9 \cdot 10^{20}$	$2.2 \cdot 10^{20}$	—	$> 2.8 \cdot 10^{19}$ AVIGNONE (1983)

So the geochemical experiments would require a nearly complete vanishing or cancellation of all ββ-amplitudes, leading through excited states in the intermediate nuclei (see Fig. 31), a situation not understood in this model. As shown already by GROTZ, KLAPDOR, METZINGER (1983 a), the ββ-rates depend quite strongly on the pairing interaction, decreasing with decreasing pairing correlations. However, to get agreement with the geochemical rates we would have to assume $\Delta_p \sim \Delta_n \lesssim 0.08$ MeV, in clear contradiction to occupation numbers deduced from stripping (MOORE et al. (1967)) and pick-up (AUBLE et al. (1968)) reactions. Also use of different sets of s.p. levels although leading to a somewhat different distribution of GT strength in the region of the first few MeV excitation energy (see Fig. 31) do not make a large effect. One might further argue that the coupling of the 2 q.p. states in the intermediate nucleus to the background of more complicated configurations could produce a statistical distribution of the phases of the matrix elements in eq. (18). However, it has to be noticed that the sum rule given by the closure matrix elements has to be fulfilled and that the major part of this sum rule strength is concentrated in the GTGR, which rules out this argument.

One has to keep in mind, however, that in the calculation of the ββ decay assumptions have been made, the validity of which has not yet been completely clarified. (This has been pointed out by KLAPDOR and GROTZ (1983) and GROTZ and KLAPDOR (1983)). One is the neglect of the distortion of the electron wave functions by the interaction of both electrons. According to a very simple energetic argument the spin-spin interaction should lead to sizeable corrections for transitions via the GTGR. The spin-spin interaction energy for two electrons separated by a distance $r = \hbar c/\Delta E$ corresponding to a virtual excitation of 10 MeV is of the same order as the Coulomb interaction with the nucleus and predominantly of opposite sign as can be seen from angular correlations.

The second is that ββ decay besides $K^0 \leftrightarrow \overline{K^0}$ strangeness oscillations is the only weak second order process and that the renormalization of a second order process is not necessarily given by $F_A{}^2$ as assumed in eq. (18) where F_A is the effective renormalized axial vector coupling constant observed in neutron decay. While $K^0 \leftrightarrow \overline{K^0}$ oscillations constitute a very delicate theoretical problem (MOHAPATRA et al. (1973)) of renormalization effects and can be understood qualitatively as a cancellation of various gauge group dependent higher order terms, in ββ decay a generalization of the phenomenological theory for first order processes seems to be straightforward.

The main point is, however, that a massive reduction of the ββ matrix elements can results from admixtures of collective quadrupole states into the g.s. wave functions of the ββ-decaying nuclei-mainly by their effect on the β^+ strength (GROTZ and KLAPDOR (1983)). The first order corrections of the ββ-decay are given by the decay of the two-phonon 0^+ states mixed with the parent ground state into p—h configurations in the intermediate nucleus and by the decay of the p—h intermediate configurations into two-phonon 0^+ states mixed with the final ground state. Table 6 gives the results obtained by KLAPDOR and GROTZ (1984).

The understanding of the absolute ββ decay rates is of particular importance also since the origin of a discrepancy between calculation and experiment might not only be sought in nuclear theory. The extremely long ββ decay half lives (up to 10^{24} a for ^{128}Te) reaching almost the order of magnitude expected for the proton ($10^{30 \pm 2}$ a) could lead one to think of an effect presently discussed (FLEMING (1983)) for the decay of the proton, in connection with tests of GUT's, namely the possibility that the samples of ββ-decaying nuclei have not yet reached the exponential stage of decay (KLAPDOR and GROTZ (1983)). Such a deviation from the exponential decay formula is quantum mechanically rigorously demanded for a time period T_0 after preparation, very short compared to the half life (FLEMING (1983), KHALFIN (1982)). The relevant parameter for assessing deviation from exponential decay is the dispersion, i.e. rms deviation ΔE, of the energy spectrum (for fixed momentum) of the parent particle (FLEMING (1983), KHALFIN (1982)). For ^{128}Te we have $T_0/\tau \cong 10^{-14}$ if we denote by τ the measured half life and if we take for T_0 the age of the universe. Geochemical experiments would measure a mean value of an effective half life τ_{eff} from sample preparation up to now, being larger than the 'intrinsic' half life corresponding to the decay rate reached after a sufficiently long time period. (This effect might be visible also as a difference between directly and geochemically measured half lives). Some further discussion can be found in KLAPDOR (1983b) and GROTZ and KLAPDOR (1983).

3.2.4. Nucleon density distributions and charge exchange reactions

The present section discusses a connection between the GT strength distribution and nuclear charge distribution. Previous, although partly rather sophisticated calculations (FIEBIG and WAMBACH (1982), SAGAWA and NGUYEN VAN GIAI (1982), MUTO et al. (1982)) of the GT strength in ^{208}Bi, including $\Delta - N^{-1}$ interaction, 2p — 2h and phonon coupling do not reproduce the observed distribution, particularly at low excitation energies.

It has been shown resently, however, (GROTZ et al. (1983)) that a rather simple model including besides the GT-residual interaction only BCS-pairing correlations can reproduce well the main features of the observed strength below the GTGR in ^{208}Bi. Previous calculations did not take into account such correlations neglecting not only the experimental evidence from the ^{208}Pb(d, t)^{207}Pb reaction for a 5% occupation of the $\nu 2g_{9/2}$ orbit (MAYER et al. (1980)), but also the fact, that the measured charge distribution (FROIS (1979)) is much flatter than that predicted by Hartree-Fock calculations (NEGELE (1970)) neglecting g.s. correlations.

It is found that g.s. correlations and thus indirectly details of the charge distribution in ^{208}Pb can influence the GT-strength function in such a way that they become observable in a high-resolution neutron time-of-flight measurement of the ^{208}Pb(p, n)^{208}Bi reaction (Fig. 32).

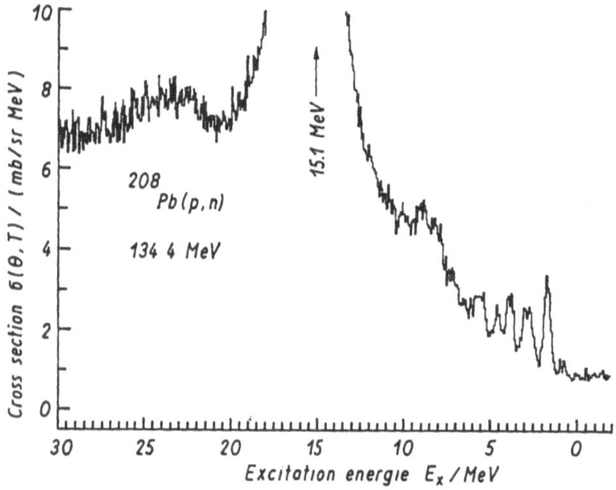

Fig. 32. Differential cross-section at 0° for the ^{208}Pb (p, n) ^{208}Bi reaction at 134.4 MeV (from GROTZ et al. (1983))

Results are shown in Fig. 33 for two values of the critically entering $\nu 1h_{11/2}$ level, positioned 0.8 (1.2) MeV below (above) the oscillator energy (*spl* I and II, respectively, in Fig. 33). Note that the strength of the transition labelled "A" in Fig. 33 is reduced markedly when going from *spl* I to *spl* II. Although transition "A" from *spl* I might be identified with the structure observed in the (p, n) reaction between 7.5 and 10 MeV (see Fig. 32), this identification is not satisfactory because the calculated transition is too strong and also because its position is too low compared to experiment, and since an increase in the energy of "A" is accompanied by a rapid increase in its strength. The results of FIEBIG and WAMBACH (1982), MUTO et al. (1982) show, that this discrepancy cannot be removed the by coupling of 2p—2h states or by $\Delta - N^{-1}$ interactions.

We turn now to the lower part of Fig. 33, where the ground state correlations are included. The part of the GT-strength below 6.5 MeV remains relatively unaltered; however, new transitions appear above 7 MeV which are comparable in strength with the transitions at lower energies. The agreement between the observed and predicted strength in the region 8—10 MeV indicates the need for the considered g.s. correlations.

It is inherent to the model, that deviations from an inert core are decreasing smoothly, when going away from the Fermi surface and as a consequence on the proton side the $\pi 3s_{1/2}$ orbit with 21% and $\pi d_{3/2}$ orbit with 14% show the biggest vacancy. The first of these values is about what is claimed (DECHARGÉ and GOGNY (1980), DÉCHARGÉ et al. (1981)) to be necessary to explain the deviation between the measured charge distribution in ^{208}Pb and the Hartree-Fock calculations without g.s. correlations, which predict a bump in the central region, in contradiction to the experiment. In an analogous way, there should be an effect of the breakup of the neutron core on the neutron density. In calculations of the neutron distribution, the filling of the $\nu 3s_{1/2}$ level plays an important role in the region of the tin isotopes (RAY and HODGSON (1979)). Also for ^{40}Ca more information about the occupation numbers is needed (RAY and HODGSON (1979)).

In conclusion, the GT-strength function can be sensitive to small occupations of high-lying neutron orbits and small vacancies on the proton side, and thus to details of the charge distribution and neutron density of the β^--decaying nucleus. This opens a new field of application of (p, n) charge exchange reactions.

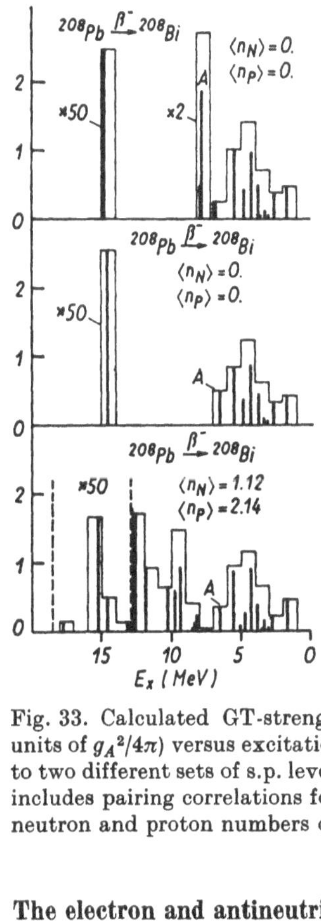

Fig. 33. Calculated GT-strength distributions (reduced β^--transition probability in units of $g_A{}^2/4\pi$) versus excitation energy in ^{208}Bi. The upper and middle parts correspond to two different sets of s.p. levels (*spl* I and II) and neglect pairing forces. The lower part includes pairing correlations for *spl.* II. $\langle n_N \rangle$ and $\langle n_p \rangle$ denote the expectation values for neutron and proton numbers outside the doubly-magic core

3.2.5. The electron and antineutrino spectra from nuclear reactors

The calculation of the electron and antineutrino spectrum in the core of a nuclear reactor presents a global test of our beta strength functions, and allows at the same time important applications. The β-spectrum is given by

$$N(E_\beta) = \sum_{Z,A} Y(Z, A, t) \cdot \sum_i b_i(Z, A, E_0{}^{(i)}) \, P_i(E_\beta, E_0{}^{(i)}, Z). \tag{19}$$

Here $Y(Z, A)$ denotes the number of beta decays per unit time of the fission fragment (Z, A) after exposure of the fissioning material to neutrons for a time t. For $t \gg \tau_\beta$ of fragment (Z, A), Y_n converges to the cumulative fission yields. The $b_i(E_0{}^{(i)})$ denotes the branching ratio of the ith beta branch in the isotope (Z, A) and $(E_0{}^{(i)})$ its beta end-point energy. $P(E_\beta, E_0{}^{(i)}, Z)$ is the normalized electron spectrum shape for end-point energy

$E_0^{(i)}$ (see, e.g. AVIGNONE and GREENWOOD (1980)). The antineutrino spectrum is obtained from the same expression (19) by replacing $P(E_\beta, E_0^{(i)}, Z)$ by $P(E_0^{(i)} - E_{\bar{\nu}}, E_0^{(i)}, Z)$.

The accuracy of the calculations of $N(E_\beta)$ or $N(E_{\bar{\nu}})$ is a direct reflection of the accuracy of the three main sets of input data to eqn. (19): The fission yields $Y(Z, A)$ (see, e.g. RIDER 1980) are relatively well-known (the uncertainties in the calculated spectra from the fission product yields are found to amount to $\lesssim 2-3\%$). The main source of uncertainty in all calculations of electron and antineutrino spectra of reactors until now were, as pointed out by KLAPDOR (1981), the assumptions on the shape of the beta strength function (which determine $b_i(E_0^{(i)})$ and $E_0^{(i)}$) for the fission products with unknown or not well-known decay schemes. Experimental decay schemes exist only for ~ 270 of the ~ 740 fission products. The β-spectra of the remaining fission products were calculated by DAVIS et al. (1979), AVIGNONE and GREENWOOD (1980), KOPEIKIN (1980) and VOGEL et al. (1981) with the assumption $S_\beta(E) \sim \rho(E)$ or with even rougher assumptions. It is thus not surprising that none of these calculations can reproduce the recent high precision measurement (SCHRECKENBACH et al. (1981)) of the β-spectrum of the fission products from ^{235}U reliably (see Fig. 34a).

Calculations using microscopic $S_\beta(E)$ for all fission products with unknown and uncertain decay schemes have been performed by KLAPDOR and METZINGER (1982, 1983). Figs. 34, 35 show the results obtained for thermal fission of ^{235}U and ^{239}Pu. Excellent agreement in absolute and relative shape with the recently measured (SCHRECKENBACH et al. (1981)) e^- spectrum from ^{235}U is obtained. Over the whole energy range important for the $\bar{\nu}$ oscillation experiments ($2.5 \leq E_{e^-} \leq 7.5$ MeV, see below) the deviation from experiment is less than 4%, i.e. about an order of magnitude smaller than in previous calculations. Our prediction of the $\bar{\nu}$ spectrum from ^{239}Pu shows excellent agreement with the recent experiment by FEILITZSCH et al. (1982).

Fig. 36 demonstrates the sensitivity of the analysis of reactor $\bar{\nu}$ oscillation experiments with respect to Δ^2 and Θ to the choice of the $\bar{\nu}$ reference spectrum for $R = 0$. The results of an analysis of the $\bar{\nu}$ spectrum measured (KWON et al. (1981)) at a distance of 8.7 m in Grenoble are shown for different $\bar{\nu}$ reference spectra. While use of the spectrum of KLAPDOR and METZINGER (1982) naturally gives a result (Fig. 36a) very similar to that

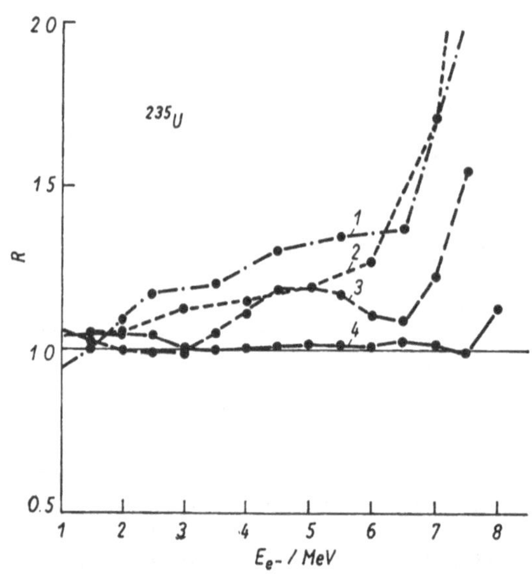

Fig. 34a

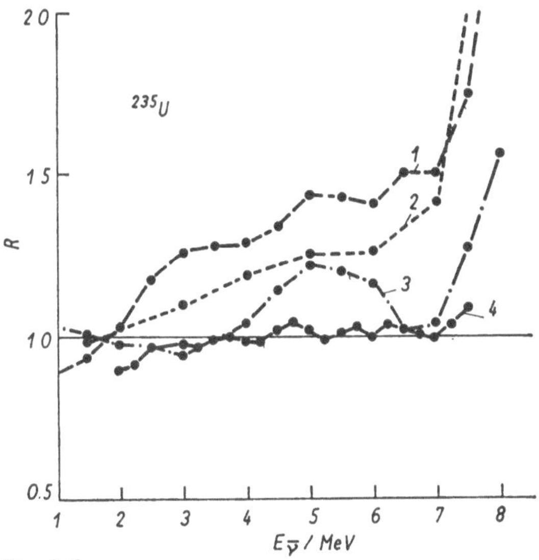

Fig. 34b

Fig. 34. (a) Calculations of the electron spectrum from thermal fission of ^{235}U. R denotes the ratio between the calculated value and the high precision experiment of SCHRECKENBACH, et al. (1981). 1–3: "Schematic" treatment of $S_\beta(E)$ by AVIGNONE and GREENWOOD (1980), KOPEIKIN (1980) and VOGEL et al. (1981), respectively. 4: microscopic calculation of $S_\beta(E)$ by KLAPDOR and METZINGER (1982a, b). (b) Calculations of the antineutrino spectrum from thermal fission of ^{235}U. R denotes the ratio between the values calculated by authors 1–4 and the calculation by KLAPDOR and METZINGER (1982a, b). 1–3: as in (a); 4: $\bar{\nu}$ spectrum extracted from the global e$^-$ spectrum by SCHRECKENBACH et al. (1981)

obtained when using the one deduced from the e$^-$ spectrum measured by SCHRECKENBACH et al. (1981), the reference spectra of AVIGNONE et al. (1980), VOGEL et al. (1981) and KOPEIKIN (1980) lead to an overestimate of the mixing angle Θ Figs. 36b, c) and thus would simulate oscillations in their absence. Consequently, calculations of the $R = 0$ $\bar{\nu}$ reference spectra with microscopic beta strength functions are indispensable for the interpretation of those reactor $\bar{\nu}$ oscillation experiments presently under way at power reactors, where the fuel composition is complex and varies as a function of time (typical average percentages of the fuel components in the Gösgen-experiment (VUILLEUMIER et al. (1982)) are $\sim 30\%$ ^{239}Pu, 7% ^{238}U and 5% ^{241}Pu). This is the more the case since there exist still not understood discrepancies in *measured* e$^-$ spectra from fission of ^{235}U and ^{239}Pu (SCHRECKENBACH et al. (1981), FEILITZSCH et al. (1982), BOROWOI et al. (1983)) and affects also (see KLAPDOR (1984a)) recent 'two-distance' measurements (GABATHULER et al. (1984)).

Finally, a comment may be allowed considering the consistency of some of the existing reactor $\bar{\nu}$ oscillation experiments. In their elegant approach of avoiding any calculated $\bar{\nu}$ reference spectrum in the analysis of $\bar{\nu}$ oscillation experiments SILVERMAN and SONI (1981a) fitted simultaneously all existing reactor $\bar{\nu}$ oscillation experiments (ccp experiments at 6.5, 8.7 and 11.2 m and ncd experiment at 11.2 m) and the shape of the $\bar{\nu}$ spectrum at $R = 0$. This latter spectrum obtained does not agree with the correct $R = 0$ reference spectrum, but is similar to that of AVIGNONE and GREENWOOD (1980) up to $E_{\bar{\nu}} = 6$ MeV, and then decreases considerably below the values of KLAPDOR and METZINGER (1982), or of SCHRECKENBACH et al. (1981). This indicates some inconsistency between the different experiments and requires further investigation.

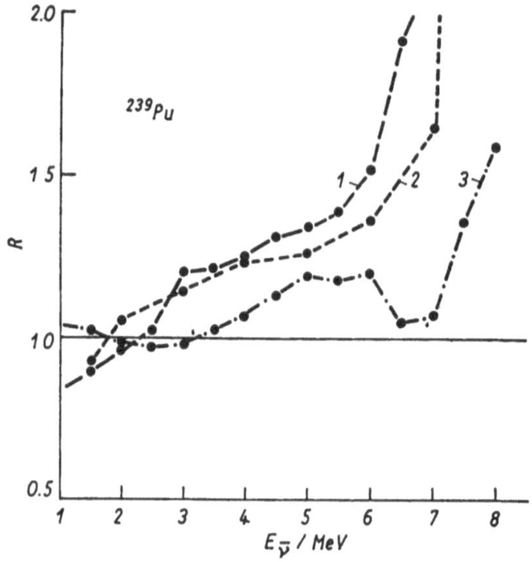

Fig. 35. Calculations of the antineutrino spectrum from thermal fission of ^{239}Pu. R denotes the ratio between the number of $\bar\nu$ per MeV and fission calculated by authors 1–3, who use a "schematic" $S_\beta(E)$ (1: Avignone and Greenwood (1980), 2: Kopeikin (1980), 3: Vogel, et al. (1981)) and the calculation by Klapdor and Metzinger (1982a) who calculated the beta strength function microscopically

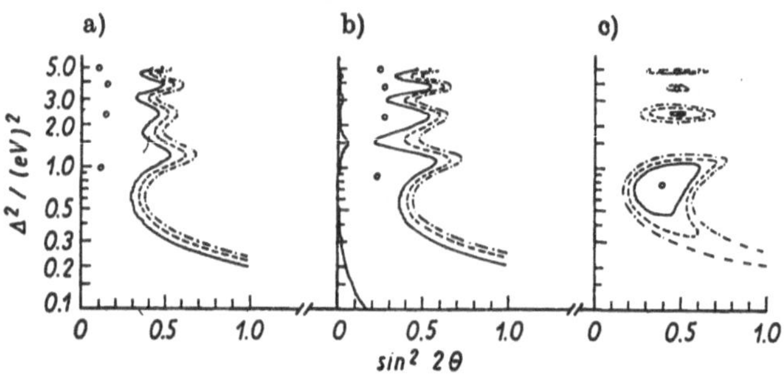

Fig. 36. Analysis of the 8.7 m ccp $\bar\nu$ oscillation experiment of Kwon et al. (1981) (under the two-neutrino hypothesis) with different core reference spectra. The relative and systematical errors given by Kwon, et al. (1981) are taken into account. Regions outside the closed areas (b, c) or right from the contour lines (a) are excluded at a confidence limit of 90% (solid line), 99% (dashed line) and 99.9% (dash-dotted line). Circles denote minima of χ^2. (a) $\bar\nu$ core spectrum from Klapdor, and Metzinger (1982a, b) — (b) from Vogel et al. (1981) — (c) from Avignone, and Greenwood (1980) (see also Fig. 34b). The spectra used in (b) and (c) lead to an overestimate of the mixing angle θ and thus would simulate oscillations in their absence (from Klapdor and Metzinger (1982b)

Acknowledgement

The author would like to thank his colleagues J. Metzinger, T. Oda, K. Grotz, F. K. Thielemann, W. Hillebrandt from the Max-Planck-Institutes and R. Madey and his coworkers from the Kent State University for the excellent cooperation over the years in which many of the reported results have been worked out. It is further a pleasure for the author to acknowledge fruitful discussions with F. Begemann, A. G. W. Cameron, F. von Feilitzsch, W. A. Fowler, G. Herrmann, H. Horie, T. Kirsten, Yu. Lazarew, Yu. Oganessian, H. Palme, F. Reines, K. Schreckenbach and P. Vogel.

References

ABERG, S., LARSSON, S. E., MØLLER, P., NILSSON, S. G., LEANDER, G., RAGNARSSON, I., Proc. Internat. Symp. Phys. and Chem. of Fission, Jülich 1979, Vol. 1 (1980), p. 303, IAEA, Vienna.

ALEKSANKIN, V. G., LYUTOSTANSKII, YU. S., PANOV, I. V., Yad. Fiz. 34 (1981) 1451.

ANDERS, E., EBIHARA, M., Geochim. Cosmochim. Acta 46 (1982) 2363.

ARNETT, W. D., Astrophys. J. 218 (1977) 815.

AUBLE, R. L., BALL, J. B., FULMER, Y. B., Nucl. Phys. A 116 (1968) 14.

AUDOUZE, J., TINSLEY, B. M., Ann. Rev. Astron. Astrophys. 14 (1976) 43.

AVIGNONE, F. T., et al., Phys. Rev. Lett. 50 (1983) 721.

AVIGNONE, F. T., GREENWOOD, Z. D., Phys. Rev. C 22 (1980) 594.

BACK, B. B., BRITT, H. C., HANSEN, O., LEROUX, B., GARRETT, J. D., Phys. Rev. C 10 (1974b) 1984.

BACK, B. B., HANSEN, O., BRITT, H. C., GARRETT, J. D., Phys. Rev. C 9 (1974a) 1924.

BAHCALL, J. N., Rev. Mod. Phys. 50 (1978) 881, and Rev. Mod. Phys. 54 (1982) 767.

BARDIN, R. K., et al., Nucl. Phys. A 158 (1970) 337.

BATIST, L. H., BERLOVICH, E., GAWRILOW, W. W., NOVIKOV, YU. N., ORLOV, C. YU., TICHONOW, W. I., Institute of Nuclear Physics, Leningrad, Preprint No. 363 (1977).

BEER, H., KÄPPELER, F., Phys. Rev. C 21 (1980) 534.

BEER, H., KÄPPELER, F., WISSHAK, K., WARD, R. A., Astrophys. J. Suppl. 46 (1981) 295.

BEGEMANN, F., Rep. Progr. Phys. 43 (1980) 1309.

BELL, G. I., Rev. Mod. Phys. 39 (1967) 59.

BENGTSSON, R., BOLEU, R., LARSSON, S. E., and RANDRUP, J., Phys. Scr. 10 A (1974) 142.

BERTSCH, G. F., The Practitioners Shell Model, North-Holland, Amsterdam (1972), p. 87.

BERTSCH, G. F., Proc. Internat. Conf. Nucl. Phys., Berkeley 1980; Nucl. Phys. A 354 (1981) 157 c.

BETHE, H. A., APPLEGATE, J. H., BROWN, G. E., Astrophys. J. 241 (1980) 343.

BLAKE, J. B., SCHRAMM, D. H., Astrophys. J. 14 (1973) 207.

BLAKE, J. B., WOOSLEY, S. E., WEAVER, T. A., SCHRAMM, D. N., Astrophys. J. 248 (1981) 315.

BOHR, A., MOTTELSON, B., Nuclear Structure, Vol. I, Benjamin, New York (1969).

BOHR, A., MOTTELSON, B., Nuclear Structure, Vol. II, Benjamin, New York (1975).

BOLEU, R., NILSSON, S. G., SHELINE, R. K., TAKAHASHI, K., Phys. Lett. 40 B (1972) 517.

BOROWOI, A. A., KLIMOW, YU. A., KOPEIKIN, W. I., Yad. Fiz. 37 (1983) 1345.

BRITT, H. C., BOLSTERLI, M., NIX, J. R., NORTON, J. L., Phys. Rev. C 7 (1973) 801.

BRITT, H. C., Proc. Internat. Symp. Phys. and Chem. Fission, Jülich 1979, Vol. 1 (1980), p. 3, IAEA, Vienna.

BROWN, G. E., Unified Theory of Nuclear Models and Forces, North-Holland, Amsterdam (1967).

BROWN, G. E., Comm. Astrophys. Space Sci. 7 (1977) 67.

BROWNE, J. C., BERMAN, B. L., Phys. Rev. C 23 (1981) 1434.

BRYMAN, D., PICCIOTTO, C., Rev. Mod. Phys. 50 (1978) 11.

BURBIDGE, E. M., BURBIDGE, G. R., FOWLER, W. A., HOYLE, F., Rev. Mod. Phys. 29 (1957) 547.

CAMERON, A. G. W., Publ. Astron. Soc. Pacific 69 (1957) 201.

CAMERON, A. G. W., Space Sci. Rev. 15 (1973) 121.

CAMERON, A. G. W., Astrophys. Space Sci. 82 (1982) 123.

CAMERON, A. G. W., COWAN, J. H., KLAPDOR, H. V., METZINGER, J., ODA, T., TRURAN, J. W., Astrophys. Space Sci. 91 (1983) 221.

CHEN, J. H., WASSERBURG, G. J., Lunar Planet Sci. XI (1980) 131.

CLEVELAND, B. T., et al., Phys. Rev. Lett. 35 (1976) 757.

COWAN, J. J., CAMERON, A. G. W., TRURAN, J. W., preprint, Univ. Oklahoma (1982).

DAVIS, R., HARMER, D. S., HOFFMANN, K. C., Phys. Rev. Lett. 20 (1968) 1205.

DAVIS, R., Proc. Conf. Status and Future Solar Neutrino Research (Ed. G. Friedlander) Vol. 1 (1978), 1, BNL 50879.

DAVIS, B. R., VOGEL, P., MANN, F. M., SCHENTER, R. E., Phys. Rev. C 19 (1979) 2259.

DECHARGE, J., GOGNY, D., Phys. Rev. C 21 (1980) 1568.

DECHARGE, J., et al., Nucl. Phys. A 358 (1981) 203 c.

DETRAZ, C., EPHERRE, M., GUILLEMAUD, D., HANSEN, P. G., JONSON, B., KLAPISCH, R., LANGEVIN, M., MATTSSON, S., NAULIN, F., NYMAN, G., POSKANZER, A. M., RAVN, H. L., DE SAINT-SIMON, M., TAKAHASHI, K., THIBAULT, C., TOUCHARD, F., Phys. Lett. 94 B (1980) 307.

Doering, R. R., Galonsky, A., Patterson, D. M., Bertsch, G. F., Phys. Rev. Lett. **35** (1975) 1691.

Doi, M., et al., Progr. Theor. Phys. **69** (1983) 602.

Dorn, D. W., Hoff, R. W., Phys. Rev. Lett. **14** (1965) 440.

Drozd, R. J., Morgan, C. J., Podosek, F. A., Poupeau, G., Shirk, J. R., and Taylor, G. J., Astrophys. J. **212** (1977) 567.

Eccles, S. F., Proc. Symp. Engineering Nuclear Explosives, Las Vegas, Nevada, Vol. **2** (1979), p. 1269.

Engler, G., Neeman, E., Nuclear Phys. **A 367** (1981) 29.

Feilitzsch, F. von, Hahn, A. A., Schreckenbach, K., Phys. Lett. **118 B** (1982) 162.

Fiebig, H. R., Wambach, J., Nucl. Phys. **A 386** (1982) 381.

Fiset, E. D., Nix, J. R., Nucl. Phys. **A 193** (1972) 647.

Fleming, G. N., Phys. Lett. **125 B** (1983) 287.

Fowler, W. A., Hoyle, F., Ann. Phys. (N.Y.) **10** (1960) 280.

Fowler, W. A., in: Cosmology, Fusion and Other Matters (Ed. F. Reines), Colorado Assoc. Univ. Press. Boulder (1972), p. 67.

Fowler, W. A., in: Proc. R. A. Welch Foundation Conf. Chem. Res. XXI Cosmochem. Houston (1978), p. 61.

Freedman, D. Z., Phys. Rev. **D 9** (1974) 1389.

Frois, B., Internat. Conf. Nucl. Phys. Electromagnetic Interactions, Mainz (1979).

Fujita, J. I., Ikeda, K., Nucl. Phys. **67** (1965) 145.

Fujita, J. I., Ikeda, K., Progr. Theor. Phys. **36** (1966) 288.

Gaarde, C., Larsen, J. S., Goodman, C. D., Foster, C. C., Goulding, C. A., Horen, D. J., Masterson, T., Rapaport, J., Tadeucci, T. N., Sugarbaker, E., Proc. 4th Internat. Conf. Nuclei far from Stability, Helsingør, CERN-Report 81-09 (1981), p. 281.

Gabathuler, K., et al., Phys. Lett. **138 B** (1984) 449.

Gangrsky, Yu. P., Marinescu, G. M., Miller, M. B., Samosyuk, V. N., Charisow, I. F., Sov. J. Nucl. Phys. **27** (1978) 475.

Gangrsky, Yu. P., Miller, M. B., Michailow, L. W., Charisow, I. F., Sov. J. Nucl. Phys. **31** (1980) 162.

Gaponow, Yu. V., Lyutostanskii, Yu. S., Fiz. elementarn. chast. i atomnovo Jadra **12** (1981) 1324.

Goodman, C. D., Goulding, C. A., Greenfield, M. B., Rapaport, J., Bainum, D. E., Foster, C. C., Love, W. G., Petrovich, F., Phys. Rev. Lett. **44** (1980) 1755.

Grotz, K., Klapdor, H. V., Metzinger, J., Madey, R., Pairsuwan, W., Anderson, B. D., Baldwin, A. R., Flanders, B., Lebo, C., Watson, J. W., Foster, C. C., Phys. Lett. **126 B** (1983) 417.

Grotz, K., Klapdor, H. V., Metzinger, J., J. Phys. **G 9** (1983a) L 169.

Grotz, K., Klapdor, H. V., Metzinger, J., Phys. Lett. **132 B** (1983b) 22.

Grotz, K., Klapdor, H. V., Report MPI H-1983-V35.

Habs, D., Klewe-Nebenius, H., Metag, V., Neumann, B., Specht, H., Z. Phys. **A 285** (1978) 53.

Hainebach, K. L., Schramm, D. N., Astrophys. J. **212** (1977) 347.

Halbleib, J. A., Sörensen, R. A., Nucl. Phys. **A 98** (1967) 542.

Halprin, A., et al., Phys. Rev. **D 13** (1976) 2567.

Hampel, W., Proc. Internat. Conf. Neutrino Phys. Astrophys., Erice, Italy 1980, Plenum Press, New York (1982), p. 61.

Hampel, W., unpublished (1980).

Haxton, W. C., Comm. Nucl. Part. Phys. **11** (1983) 41.

Haxton, W. C., Stephenson, G. J., Strottman, D., Phys. Rev. Lett. **47** (1981) 153.

Haxton, W. C., Stephenson, G. J., Strottman, D., Phys. Rev. **D 25** (1982) 2360.

Hennecke, E. W., Manuel, O. K., Sabu, D. D., Phys. Rev. **C 11** (1975) 1378.

Hilf, E. R., von Grote, H., Takahashi, K., Proc. 3rd Internat. Conf. Nuclei far from Stability, CERN-Report 76-13 (1976), p. 142.

Hillebrandt, W., Space Sci. Rev. **21** (1978) 639.

Hillebrandt, W., Proc. 4th EPS Gen. Conf. (1979), p. 255.

Hillebrandt, W., Proc. Nato Adv. Study Supernovae, Cambridge, England, June 29—July 11 (1981).

HILLEBRANDT, W., Phys. Bl. 38 (1982) 189.
HILLEBRANDT, W., TAKAHASHI, K., KODAMA, T., Astron. Astrophys. **52** (1976) 63.
HILLEBRANDT, W., THIELEMANN, F. K., Mitt. Astron. Ges. **43** (1977) 234.
HOFF, R. W., Proc. Symp. 25th Anniv. Elements 99 and 100, Berkeley, 1978, Preprint UCRL-81566, Lawrence Livermore Lab. (1978).
HOFF, R. W., HULET, E. K., Proc. Symp. Engineering Nucl. Explosives, Las Vegas, Nevada, Vol. **2** (1970), p. 1283 (HUTCH).
HOFFMAN, D. C., Ark. Fys. **36** (1967) 533 (CYCLAMEN).
HOWARD, W. M., MØLLER, P., At. Data Nucl. Data Tables **25** (1980) 219.
HOWARD, W. M., NIX, J. R., Nature **247** (1974) 17.
HUDSON, B., HOHENBERG, C. M., KENNEDY, B. M., PODOSEK, F. A., Lunar Planet. Sci. **XIII** (1982) 346.
HUFFMAN, A. H., Phys. Rev. **C 2** (1970) 742.
HULET, E. K., WILD, J. F., LOUGHEED, R. W., EVANS, J. E., QUALHEIM, B. J., NURMIA, M., GHIORSO, A., Phys. Rev. Lett. **26** (1971) 523.
HULET, E. K., ECCLES, S. F., HECKMAN, R. A., unpublished (1973).
IBEN, I., Science **155** (1968) 785.
IBEN, I., Ann. Rev. Astron. Astrophys. **12** (1974) 215.
INGLEY, J. S., Nucl. Phys. **A 124** (1969) 130 (summary of all explosions up to 1969).
JÄNECKE, J., At. Data Nucl. Data Tables **17** (1976) 455.
JEUKENNE, J. P., LEJEUNE, A., MAHAUX, D., Phys. Rev. **C 182** (1977) 1190.
JOHN, W., HULET, E. K., LOUGHEED, R. W., WESOLOWSKI, J. J., Phys. Rev. Lett. **27** (1971) 45.
JORDAN, J., KIRSTEN, RICHTER, H., Z. Naturforsch. **35a** (1980) 145.
KÄPPELER, F., BEER, H., WISSHAK, K., CLAYTON, D. D., MACKLIN, R. L., WARD, R. A., Astrophys. J. **257** (1982) 821.
KHALFIN, L. A., Phys. Lett. **112 B** (1982) 223.
KIRSTEN, T., in Proc. Workshop Sci. Underground, Los Alamos 1982.
KIRSTEN, T., MÜLLER, H. W., Earth Planet Sci. Lett. **6** (1969) 271.
KIRSTEN, T., RICHTER, H., JESSBERGER, E., Phys. Rev. Lett. **50** (1983) 474.
KIRSTEN, T., RICHTER, H., JESSBERGER, E., Z. Phys. **C 16** (1983) 189.
KLAPDOR, H. V., Phys. Lett. **35 B** (1971) 405.
KLAPDOR, H. V., Phys. Lett. **65 B** (1976) 35.
KLAPDOR, H. V., Phys. Rev. **C 23** (1981) 126.
KLAPDOR, H. V., Physik Bl. 38 (1982) 182.
KLAPDOR, H. V., Progr. Part. Nucl. Phys. **10** (1983a) 131.
KLAPDOR, H. V., Proc. Internat. Symp. Electromagn. Prop. Atomic Nuclei, Tokyo, 9–12 November 1983, Preprint MPI H-1983-V28 (1983b).
KLAPDOR, H. V., Proc. Internat. Symp. on Nuclear Spectroscopy and Nuclear Interactions, Osaka, Japan, 21–24 March 1984a.
KLAPDOR, H. V., Sterne und Weltraum, in press (1984).
KLAPDOR, H. V., GROTZ, K., Report MPI H-1983-V34 (1983).
KLAPDOR, H. V., GROTZ, K., Phys. Lett. **142 B** (1984) 323.
KLAPDOR, H. V., METZINGER, J., Phys. Rev. Lett. **48** (1982a) 127.
KLAPDOR, H. V., METZINGER, J., Phys. Lett. **112 B** (1982b) 22.
KLAPDOR, H. V., METZINGER, J., Internal Rep. (1982c).
KLAPDOR, H. V., METZINGER, J., Proc. 8th Internat. Workshop on Weak Interactions and Neutrinos, Javea (Alicante), Spain, 6–11 Sept. 1982, p. 87.
KLAPDOR, H. V., ODA, T., Astrophys. J. **242** (1980) L49.
KLAPDOR, H. V., WENE, C. O., Astrophys. J. **230** (1979) L113.
KLAPDOR, H. V., WENE, C. O., J. Phys. **G 6** (1980) 1061.
KLAPDOR, H. V., METZINGER, J., ODA, T., Report MPI H-1981-V47, partly publ. in: Chart of the Nuclides, 5. edition, Karlsruhe, Nov. 1981 (1981c); Z. Phys. **A 309** (1982) 91.
KLAPDOR, H. V., METZINGER, J., ODA, T., Atomic Data Nucl. Data Tables, 31, No. 1 (1984) 81.
KLAPDOR, H. V., WENE, C. O., ISOSOMOW, I. N., NAUMOW, YU. W., Phys. Lett. **78 B** (1978) 20.
KLAPDOR, H. V., WENE, C. O., ISOSIMOW, I. N., NAUMOW, YU. W., Z. Phys. **A 292** (1979) 249.
KLAPDOR, H. V., ODA, T., METZINGER, J., HILLEBRANDT, W., THIELEMANN, F. K., Z. Phys. **A 299** (1981a) 312.

Klapdor, H. V., Oda, T., Metzinger, J., Thielemann, F. K., Hillebrandt, W., Proc. 4th Internat. Conf. Nuclei far from Stability, Helsingør, CERN-Report 81-09 (1981b), p. 341.

Kodama, T., Takahashi, K., Phys. Lett. **43 B** (1973) 167.

Kodama, T., Takahashi, K., Nucl. Phys. **A 239** (1975) 489.

Kopeikin, W. I., Yad. Fiz. **32** (1980) 1507.

Krumlinde, J., Møller, P., Wene, C. O., Howard, W. M., CERN-Report 81-09 (1981), p. 260.

Kusnetsov, V. I., Skobelev, N. K., Flerov, G. N., Yad. Fiz. **4** (1966) 279.

Kwon et al., Phys. Rev. **D 24** (1981) 1097.

Lattimer, J. M., Ann. Rev. Nucl. Part. Sci. **31** (1981) 337.

Lazarew, Yu. A., Oganessian, Yu. Ts., Kusnetsov, V. I., Internat. Symp. Synthesis and Properties New Elements, Dubna, 23–27 Sept. 1980, preprint Dubna E7-80-719 (1980).

Los Alamos Radiochemistry Group, Phys. Rev. Lett. **14** (1965) 962 (BARBEL).

Lund, E., Hoff, P., Alekett, K., Glomset, O., Rudstam, G., Z. Phys. **A 294** (1980) 233.

Lynn, J. E., in: Nuclear Theory for Applications, Proc. Internat. Center Theoret. Physics, Trieste 1978, IAEA, Vienna, 1980.

Marti, K., Lugmair, G. W., Scheinin, N. B., Lunar Sci. **VIII** (1977) 619.

Mayer, R. A., Cohen, B. L., Diehl, R. C., Phys. Rev. **C 2** (1970) 1898.

Meldner, H., Phys. Scr. **10 A** (1974) 149.

Meldner, H. W., Cowan, G. A., Nix, J. R., Stoughton, R. W., Phys. Rev. **C 13** (1976) 182.

Metag, V., Habs, D., Specht, H. J., Phys. Rep. **65** (1980) 1.

Metzinger, J., Klapdor, H. V., Annual Report MPI Kernphysik (1983).

Miller, G. E., Scalo, J. M., Astrophys. J. Suppl. **41** (1979) 513.

Minkowski, P., Nucl. Phys. **B 201** (1982) 269.

Moe, M. K., Lowenthal, D. D., Phys. Rev. **C 22** (1980) 2186.

Mohapatra, R. N., Pati, J. C., Vinciarelli, P., Phys. Rev. **D 8** (1973) 3652.

Møller, P., Nix, J. R., Proc. 3rd Symp. Physics and Chem. Fission, Rochester, Vol. 1 (1973), p. 102, IAEA, Vienna.

Møller, P., Nix, J. R., Nucl. Phys. **A 229** (1974) 269.

Møller, P., Nix, J. R., At. Data Nucl. Data Tables **26** (1981) 165.

Møller, P., Nucl. Phys. **A 192** (1975) 529.

Møller, P., Proc. Internat. Symp. Phys. and Chem. Fission, Jülich 1979, Vol. **1** (1979), p. 283, IAEA, Vienna.

Moore, W. H., Schlegel, G. K., O'Dell, S., Nucl. Phys. **A 104** (1967) 327.

Muto, K., Oda, T., Horie, H., Phys. Lett. **118 B** (1982) 261.

Myers, W. D., Droplet Model Atomic Nuclei, Plenum Press, New York (1977).

Nissen, P. E., ESO Messenger **28** (1982) 4.

Norman, E. B., Schramm, D. N., Astrophys. J. **228** (1979) 881.

Oda, T., private communication (1981).

Ohnishi, T., Technical Report of the Institute of Atomic Energy, Kyoto University No. 179 (1979).

Pagel, B. E. J., Edmunds, M. G., Ann. Rev. Astron. Astrophys. **19** (1981) 77.

Pauli, H. C., Ledergerber, T., Nucl. Phys. **A 175** (1971) 545.

Podosek, F. A., Geochim. Cosmochim. Acta **36** (1972), 755.

Pontecorvo, B., Phys. Lett. **26 B** (1968) 630.

Priester, W. preprint (1984).

Primakoff, H., Rosen, S. P., Rep. Prog. Phys. **22** (1959), Ann. Rev. Nucl. Part. Sci. **31** (1981) 145.

Ragnarsson, I., Proc. Internat. Conf. Nuclei far from Stability, Leysin, CERN-Report 70-30 (1970), p. 847.

Ragnarsson, I., in: Future Directions in Studies of Nuclei far from Stability (Ed. J. H. Hamilton), North-Holland, Amsterdam (1980), p. 367.

Randrup, J., Nucl. Phys. **A 207** (1973) 209.

Randrup, J., Larsson, D. E., Møller, P., Sobiczewski, A., Lukasiak, A., Phys. Scr. **10 A** (1974) 60.

Ray, L., Hadgson, P. E., Phys. Rev. **C 20** (1979) 2403.

Reeder, P. L., Wright, J. F., Alquist, L. J., Phys. Rev. **C 15** (1977) 2108.

Reeder, P. L., Warner, R. A., Yeh, T. R., Chrien, R. E., Gill, R. L., Liou, H., Shmid, M., Shelts, M. L., CERN-Report 81-09 (1981), p. 276.

Rider, B. F., Compilation of Fission Product Yields, NEDO-12154-3(B), Vallecitos Nuclear Data Center (1980).

RISTORI, C., CRANCON, J., WÜNSCH, K. D., JUNG, G., DECKER, R., KRATZ, K.-L., Z. Phys. **A 290** (1979) 311.
ROBERTS, A., BLOOD, H., LEARNED, J., REINES, F., Proc. Int. Neutrino Conf., Aachen (1977), p. 688.
ROSEN, S. P., Proc. Internat. Conf. Neutrino Phys. and Astrophysics, Maui, Hawaii 1981, Vol. 2 (1981), p. 76.
SAGAWA, H., NGUYEN VAN GIAI. Phys. Lett. **113 B** (1982) 119.
SALPETER, E. E., SHAPIRO, S. L., Astrophys. J. **251** (1981) 311.
SANDAGE, A., TAMMANN, G. A., Astrophys. J. **256** (1982) 339.
SANDAGE, A., Astrophys. J. **252** (1982) 553.
SCHRAMM, D. N., Ann. Rev. Astron. Astrophys. **12** (1974) 303.
SCHRAMM, D. N., Ann. New York Acad. Sci. **336** (1980) 380.
SCHRAMM, D. N., FOWLER, W. A., Nature **231** (1971) 103.
SCHRAMM, D. N., FISET, E. D., Astrophys. J. **180** (1973) 551.
SCHRECKENBACH, K., FAUST, H. R., FEILITZSCH, F. VON, HAHN, A. A., HAVERKAMP, K., VUILLEUMIER, J. L., Phys. Lett. **99 B** (1981) 251.
SEEGER, P. A., FOWLER, W. A., CLAYTON, D. D., Astrophys. J. Suppl. Ser. **11** (1965) 121.
SEEGER, P. A., SCHRAMM, D. N., Astrophys. J. **160** (1970) L157.
SILVERMAN, D., SONI, A., Phys. Rev. Lett. **46** (1981) 467.
SRINIVASAN, B., et al., Econ. Geol. **68** (1973) 252.
SRINIVASAN, B., ALEXANDER, B. C., MANUEL, O. K., Econ. Geol. **67** (1972) 592.
SYMBALISTY, E. M. D., SCHRAMM, D. N., Rep. Progr. Phys. **44** (1981) 293.
TAKAHASHI, K., YAMADA, M., KONDOH, Z., At. Data Nucl. Data Tables **12** (1973) 101.
THIELEMANN, F. K., METZINGER, J., KLAPDOR, H. V., Z. Phys. **A 309** (1983) 301, and Astron. Astrophys. **123** (1983) 162.
TINSLEY, B. M., Astrophys. J. **198** (1975) 145.
TINSLEY, B. M., Fund. Cosmic Phys. **5** (1980) 287.
TRIMBLE, V., Rev. Mod. Phys. **47** (1975) 877.
TRIMBLE, V., Rev. Mod. Phys. **54** (1982) 1183 and **55** (1983) 511.
TRIVEDI, B. M. P., Astrophys. J. **215** (1977) 877.
TRURAN, J. W., COWAN, J. J., CAMERON, A. G. W., Astrophys. J. **222** (1978) L63.
TUBBS, D. L., SCHRAMM, D. N., Astrophys. J. **201** (1975) 467.
VIOLA, V. E., SWANT, J. A., GRABER, J., At. Data Nucl. Data Tables **13** (1974) 35.
VOGEL, P., SCHENTER, G. K., MANN, F. M., SCHENTER, R. E., Phys. Rev. **C 24** (1981) 1543.
VUILLEUMIER, J. L., BOEHM, F., EGGER, J., FEILITZSCH, F. VON, GABATHULER, K., GIMLETT, J. L., HAHN, A. A., KWON, H., MÖSSBAUER, R. L., ZACEK, G., ZACEK, V., Phys. Lett. **114 B** (1982) 298.
WARD, R. A., CLAYTON, D. D., priv. communication and to be publ. (1982).
WASSERBURG, G. J., HUNECKE, J. C., BURNETT, D. S., J. Geophys. Res. **74** (1969) 4221.
WENE, C. O., Astron. Astrophys. **44** (1975) 233.
WENE, C. O., JOHANSSON, S. A. E., Phys. Scr. **10 A** (1974) 156.
WENE, C. O., JOHANSSON, S. A. E., Proc. 3rd Internat. Conf. Nucl. far from Stability, Cargese, Corsica, CERN-Report 76-13 (1976), p. 581.
WILD, J. F., HULET, E. K., LOUGHEED, R. W., J. Inorg. Nucl. Chem. **35** (1973) 1063.
WILSON, J. R., Astrophys. J. **163** (1971) 209.
WILSON, J. R., Ann. New York Acad. Sci. 336 (1980) 358.
WINTERS, R. R., MACKLIN, R. L., Phys. Rev. **C 25** (1982) 208.
WOOSLEY, S. E., WEAVER, T. A., Proc. NATO Advanced Study Inst., Cambridge (1981), p. 79.
YOKOI, K., TAKAHASHI, K., ARNOLD, M., Astron. Astrophys. **117** (1983) 65.
ZAMICK, L., AUERBACH, N., Phys. Rev. **C 26** (1982) 2185.

Prof. Dr. Hans Volker Klapdor
Max-Planck-Institut für Kernphysik
D 6900 Heidelberg 1, Postfach 103980

Systems Analysis – Modelling – Simulation

Journal of Mathematical Modelling and Simulation in Systems Analysis

Edited at the Central institute of Cybernetics and Information Processes of the Academy of Sciences of the G.D.R. in cooperation with the Mathematical Society of the G.D.R. by Achim Sydow (Berlin) with the assistance of an international team of scientists

Published yearly, 1 volume (4 numbers).
Volume 1, 1984.
16,7 cm by 24,0 cm. Annual list of contents.
Annual subscription M 120.—; single number M 30.—.
Order number: 1097 *ISSN 0232-9298*

The journal is devoted to art and techniques and applications of modelling and simulation in systems analysis. Systems analysis is an interdisciplinary activity. Articles will be published on systems analysis and systems theory, modelling and mathematical models and tools for decision and control, computer simulation systems, modelling and simulation in environmental protection, macroeconomics, agricultural production, regional planning, biosciences, traffic control, large engineering systems, resources distribution, management systems etc. In addition to original papers, survey articles and reports on conferences etc. are published. Finally, books covering the above mentioned and related subjects are reviewed.

Of interest to: Scientists of systems sciences, informatics, control engineering, applied mathematics, operation research, biology, economy, physics, environmental protection, agriculture etc.

Orders can be sent

- in the GDR: to the Deutsche Post, Zentralvertrieb des PZV (B), 7930 Herzberg/Elster or to the Akademie-Verlag, DDR-1086 Berlin, Leipziger Str. 3—4, PF-Nr. 1233;
- in the other socialist countries: to a book-shop for foreign language literature or to the competent news-distributing agency;
- in the FRG and Berlin (West): to the wholesale distributing agency Kunst und Wissen, Erich Bieber OHG, Wilhelmstr. 4—6, D-7000 Stuttgart 1;
- in the other Western European countries: to Kunst und Wissen, Erich Bieber GmbH, Dufourstr. 51, CH-8008 Zürich;
- in other countries: to the Akademie-Verlag, DDR-1086 Berlin, Leipziger Str. 3—4, PF-Nr. 1233.

Akademie-Verlag · Berlin

DDR-1086 Berlin, Leipziger Straße 3—4

The journal "Fortschritte der Physik/Progress of Physics" is devoted to the theoretical and experimental study of the fundamental constituents of matter and their interactions, so to elementary particle physics, classical and quantum field theory, theory of graviation, thermodynamics and statistics, nuclear physics, laser physics, and plasma physics.
The articles should have in general review character.

Manuscript should be addressed to the Editor-in-Chief of the journal

Prof. Dr. F. Kaschluhn
Sektion Physik der Humboldt-Universität zu Berlin
DDR-1086 Berlin, PSF 1297

or to editors or to members of the Advisory Board.

Terms of subscription for the journal "Fortschritte der Physik/Process of Physics"

Orders can be sent

— *in the GDR:* to the Deutsche Post, Zentralvertrieb des PZV (B), 7930 Herzberg/Elster, or to the Akademie-Verlag, DDR-1086 Berlin, Leipziger Str. 3—4, PF-Nr. 1233;

— *in the other socialist countries:* to a book-shop for foreign language literature or to the competent news-distributing agency;

— *in the FRG and Berlin (West):* to a book-shop or to the wholesale distributing agency Kunst und Wissen, Erich Bieber OHG, Wilhelmstr. 4—6, D-7000 Stuttgart 1;

— *in the other Western European Countries:* to Kunst und Wissen, Erich Bieber GmbH, Dufourstr. 51, CH-8008 Zürich;

— *in other countries:* to the international book- and journal-selling trade, to Buchexport, Volkseigener Außenhandelsbetrieb der DDR, Postfach 160, DDR-7010 Leipzig, or to the Akademie-Verlag, DDR-1086 Berlin, Leipziger Str. 3—4, PF-Nr. 1233.

Zeitschrift „Fortschritte der Physik"

Herausgeber: Prof. Dr. F. Kaschluhn, Prof. A. Lösche, Prof. Dr. R. Rompe.
Verlag: Akademie-Verlag, DDR-1086 Berlin, Leipziger Str. 3—4;
Fernruf: 2236221 und 2236229; Telex-Nr.: 114420; Bank: Staatsbank der DDR, Berlin, Konto-Nr.: 6836-26-20712.

Chefredakteur: Dr. Lutz Rothkirch
Anschrift der Redaktion: Sektion Physik der Humboldt-Universität zu Berlin, DDR-1040 Berlin, Hessische Straße 2.

Veröffentlicht unter der Lizenznummer 1324 des Presseamtes beim Vorsitzenden des Ministerrates der Deutschen Demokratischen Republik.

Gesamtherstellung: VEB Druckhaus „Maxim Gorki".
DDR-7400 Altenburg, Carl-von-Ossietzky-Straße 30/31.

Erscheinungsweise: Die Zeitschrift „Fortschritte der Physik" erscheint monatlich. Die 12 Hefte eines Jahres bilden einen Band. Bezugspreis je Band 216,— M zuzüglich Versandspesen. Preis je Heft 18,— M.

Bestellnummer dieses Heftes: 1027/33/1.

AN (EDV) 57618

Contents of the following issues:

R. C. Edgar: Being Discrete about Yang and Mills: Basic Techniques of Euclidean Lattice Gauge Theory

L. L. DeRaad, Jr.: Source Theory Treatment of the Casimir Effect: A Review

A. W. Hendry, and D. B. Lichtenberg: Properties of Hadrons in the Quark Model

A. O. Barut, S. Komy: Derivation of Nonperturbative Relativistic Two-Body Equations from the Action Principle in Quantumelectrodynamics

A. O. Barut, Nuri Ünal: Radial Equations for the Relativistic Two-Fermion Problem with the Most General Electric and Magnetic Potentials

Ya. Z. Darbaidze, A. N. Sissakian, L. A. Slepchenko, H. T. Torosian: Analysis of Experimental Semi-Inclusive Distributions at the SPS Energies

A. Gonzáles: Radiative Corrections to Bose Condensation

M. Nishioka: Weyl's Gauge Field and Its Behaviour

H. Satz: The Limits of Hadron Thermodynamics

Progress of Physics is indexed
in Current Contents/Physical, Chemical & Earth Sciences.

FSC
www.fsc.org
MIX
Papier | Fördert
gute Waldnutzung
FSC® C083411

Zeitfracht Medien GmbH
Ferdinand-Jühlke-Straße 7
99095 Erfurt, Deutschland
produktsicherheit@kolibri360.de